Celebrating the Existence of the Most High through Equations

by Tyrone James

Wealthy Sistas Publishing House
ISBN: 978-0-9833806-8-9

All rights reserved.
No part of this publication may be reproduced, distributed, or transmitted in any form or by any means, including photocopying, recording, or other electronic or mechanical methods, without the prior written permission of the publisher, except in the case of brief quotations used in critical reviews and certain other noncommercial uses permitted by copyright law.

© 2025 Tyrone James
First Edition
Printed in the United States of America
For more information:
TyroneJames.com
WealthySistas.com
Wealthy Sistas Publishing House is a division of Wealthy Sistas Media Group, LLC.

Celebrating the
Existence of the

Most High

Through Equations

$$E(t\gamma) = \lim_{\gamma \to 0} \left| \frac{1}{1 - ke^{-\gamma t}} \right|$$

TYRONE JAMES

TABLE OF CONTENTS

Dedication vii
Foreword – Bishop Lee Anthony Norwood viii
Opening Prayer x
Preface xi
Introduction xiii
The Purpose of This Exploration
Science, Spirituality, and the Language of Mathematics

Chapter One: The Divine Architecture of the Universe 1
The Mathematical Precision of Creation
The Fibonacci Sequence & the Golden Ratio
Fractals & Sacred Geometry as Divine Signatures

Chapter Two: The Mathematics of Human Life and Consciousness 5
The Circadian Rhythms & the Divine Order
The Numerical Patterns in Human Biology
Ethical Structures & the Mathematical Design of Morality

Chapter Three: The Symphony of Universal Laws 9
The 12 Universal Laws and Their Scientific Counterparts
The Law of Attraction & The Law of Gravity
The Law of Cause and Effect & Newton's Third Law
The Law of Compensation & The Conservation of Energy
The Law of Correspondence & The Fractal Nature of Reality

TABLE OF CONTENTS (continued)

Chapter Three, Continued — 9
The Law of Divine Oneness & Quantum Entanglement
The Law of Gender & The Principle of Complementarity
The Law of Inspired Action & The Principle of Motion
The Law of Perpetual Transmutation of Energy & The Laws of Thermodynamics
The Law of Polarity & The Expansion and The Contraction of the Universe
The Law of Relativity & Einstein's Theory of Relativity
The Law of Rhythm & The Cyclical Nature of the Cosmos
The Law of Vibration & The Quantum Field

Chapter Four: Equations as Reflections of the Divine Mind — 57
Mathematical Constants in Sacred Texts
The Language of the Cosmos: The Most High's Signature in Numbers

Chapter Five: The Hidden Dimensions of Reality — 68
Multidimensional Existence in Science & Spirituality
Parallel Universes & the Unseen Realms

Chapter Six: The Fabric of the Unseen — 75
Dark Matter, Energy, and the Divine Presence
Spiritual Realms Beyond the Physical World

Chapter Seven: The Equation of Balance: Harmony in the Cosmos and Within — 82
The Fine-Tuning of the Universe
The Interconnectedness of Creation

TABLE OF CONTENTS (continued)

Chapter Eight: The Equation of Time: 89
Cycles, Seasons, and Divine Order
Time as a Created Dimension
The Eternal Nature of the Divine and the Temporal World

Chapter Nine: Conclusion & Reflection   96
The Mathematical Testament to the Divine
A Final Word on the Interplay Between Science and Faith

Closing Benediction	99
Afterword – Sheik Ahmad Hamza Abdullah	100
Appendix: Visual Diagrams & Reflections	102
About the Author	121
Notes	122

DEDICATION

This book is dedicated to the three extraordinary women who have shaped my life:

To **Kimberly**, my beloved wife in this life and the next—your spirit continues to be my most profound inspiration. You were my muse, my greatest love, and the heartbeat behind this work.

To **Cheryl**, my dearest friend and unwavering supporter, thank you for standing by me through every season. I will love you for the rest of my days.

To **Aunt Carolyn**, the steady rock of our family, your love and support have grounded me. You are a blessing beyond measure.

FOREWORD

In an age where many still perceive science and faith as opposing forces, Celebrating the Existence of the Most High through Equations offers a bold, necessary bridge—one built on reverence, logic, and the beauty of divine order.
I have long held the conviction that our Creator welcomes our inquiry—He has given us minds to reason, to explore, and to marvel. Scripture tells us, "The heavens declare the glory of God; the skies proclaim the work of his hands" (Psalm 19:1).

This book is an invitation into that declaration—where mathematics is not cold or sterile, but radiant with meaning. Where spiritual truths echo in the patterns of nature. And where the sacred texts of Judaism, Christianity, and Islam converge with physics to remind us that all knowledge ultimately points back to the Source.

Tyrone James has offered us more than a book. He's offered a lens—a way of seeing the divine fingerprints etched across the universe. Whether you are a person of deep faith, scientific curiosity, or both, I believe this work will challenge, enlighten, and uplift you.

May we never fear where truth leads, for the Most High is both the Author of Scripture and the Architect of the stars.

Bishop, Lee Anthony Norwood
High Praise Christian Ministries
Atlanta, GA

OPENING PRAYER

O Most High, Source of All that Exists—
We begin this journey in reverence and awe of Your magnificent design.

You, who hung the stars in calculated balance and coded the cosmos with sacred patterns—
We seek not just knowledge, but wisdom.

May every page written here reflect a fragment of Your truth.
May the equations and insights presented draw us nearer to understanding Your will and wonder.

Bless the reader with clarity, the seeker with guidance, and the skeptic with open possibility.

Let this work stand as a humble offering—
An expression of devotion through reason, faith, and inquiry.

In Your Name—the Eternal Architect—do we begin.
Ameen.

PREFACE

This book is the result of a lifelong journey—a pursuit to understand the invisible patterns that govern our existence, the laws that uphold creation, and the Divine Intelligence that breathes purpose into it all. As a man shaped by faith, study, and observation, I have always believed that Science, in many ways, reflects the Divine—a lens through which we glimpse the sacred, and a path to spirituality.

Throughout history, humanity has sought to describe the indescribable. We've carved meaning into stone, captured it in scripture, and traced it in the stars. And while language may differ—scientific, religious, or poetic—the desire is the same: to honor the Source from which we came.

In these pages, you will find mathematical insights, sacred references, and metaphysical reflections, not as academic conclusions but as sincere explorations. My aim is not to instruct, but to inspire; not to preach, but to provoke thought and deeper connection to the Most High.

This work exists at the intersection of faith and function, reason and revelation. It reflects my deep respect for the truths found in scripture and the wonders revealed through creation. I believe that behind every equation lies a truth, and behind every truth, a trace of the Divine.

While the three holy texts—the Quran, the Bible, and the Torah—each serve as guiding lights for billions around the world, this work draws from their shared themes of divine order, purpose, and moral law. The references are offered not as theological declarations, but as reflections on the interconnected wisdom found across these sacred traditions.

May this book be a window—not a wall. And may it bring clarity, curiosity, and reverence to all who read it.

INTRODUCTION

In the vast expanse of the universe, woven into the very fabric of existence, lies an undeniable order—a rhythm, a pattern, a set of principles that govern all things. This order is not random; it is precise, intricate, and awe-inspiring. At the heart of this divine architecture is mathematics, the universal language through which the cosmos communicates its secrets. This book is an exploration of how equations, those simple strings of numbers and symbols, reveal the profound truth of the Most High's existence.

Equations are more than just tools for scientists and mathematicians; they are reflections of a greater reality. They unveil the symmetry in nature, the harmony in motion, and the balance in forces that sustain life. Through the lens of mathematics, we glimpse the fingerprints of the Divine, etched into the core of everything from the tiniest particle to the farthest galaxy.

As we embark on this journey, it is essential to acknowledge the foundational inspiration behind this work, the profound insights, and original equations of Tyrone James. His passion for uncovering the sacred within the scientific has illuminated many of the concepts explored throughout this book. Tyrone's equations serve as bridges between the abstract world of mathematics and the tangible reality of divine

presence, guiding us to recognize the Most High in both the seen and unseen.

This book is not merely an intellectual exploration; it is a heartfelt celebration of the Most High, whose presence is reflected in every equation, every pattern, and every law of the universe. Mathematics is more than numbers and symbols; it is the language through which the Most High communicates the structure and harmony of creation. From the spirals of galaxies to the intricate design of a single leaf, mathematical principles reveal a universe woven with divine precision.

Throughout history, scholars and spiritual seekers have searched for the intersection of science and faith. This book seeks to bridge that gap, demonstrating how mathematics, far from being a mere human invention, is a divine language—a code written by the Most High into the fabric of the universe. Tyrone's original work laid the foundation for this exploration, revealing that every equation, every pattern, and every structure in existence reflects the wisdom and presence of the Creator.

In *Celebrating the Existence of the Most High through Equations*, we will journey through both the foundational equations presented by Tyrone James and new insights that elevate our understanding. This book aims to bridge the gap between scientific discovery and spiritual awakening, showing that faith and reason are not opposing forces but complementary paths to truth. We will explore the

mathematical patterns found in sacred texts, the laws that govern the cosmos, and the unseen dimensions that shape our spiritual experiences. Along the way, we will see how the Most High's signature is embedded in the very equations that describe our reality.

This is not just a book about mathematics; it is a call to reflect, to wonder, and to celebrate the divine intelligence that sustains all things. May this journey inspire a deeper connection with the Most High and a renewed sense of awe for the mathematical beauty that surrounds us.

CHAPTER ONE
The Divine Architecture of the Universe

The universe is a grand architectural masterpiece, woven together with precision and harmony by the Most High. From the spirals of galaxies to the Fibonacci sequences in nature, the mathematical order of creation reveals an undeniable intelligence beyond human comprehension.

One of the most profound reflections of divine architecture can be seen in the Golden Ratio (φ), an irrational number approximately equal to 1.618. This ratio appears throughout nature, art, and even the human body. It governs the arrangement of petals in flowers, the spiral of seashells, and the proportions of the human face. The Quran hints at this divine order: "[Allah] has created everything in proportion and measure" (Quran 54:49).

Similarly, the Bible references divine order in creation: "God saw all that he had made, and it was very good" (Genesis 1:31), emphasizing that creation follows a perfect, intentional design. The Torah echoes this concept in Exodus 26, where the precise measurements of the Tabernacle are divinely prescribed, demonstrating how sacred structures mirror the mathematical precision of the universe.

The Golden Ratio also appears in the architecture of historical places of worship, such as Solomon's Temple, reflecting divine harmony in sacred spaces. Across these scriptures, we find that mathematics is not just a scientific discovery but a spiritual language, affirming that the Most High has imbued creation with proportion, balance, and beauty.

The placement of celestial bodies further underscores this divine precision. The Earth orbits the sun at a distance perfectly suited for sustaining life—a balance so fine that even a slight shift would render the planet uninhabitable. This precise tuning is often referred to as the Anthropic Principle, a concept supported by many physicists and cosmologists. According to Stephen Hawking, "The laws of science, as we know them at present, contain many fundamental numbers... The remarkable fact is that the values of these numbers seem to have been very finely adjusted to make possible the development of life." The Earth's position within the habitable zone, the delicate balance of gravitational forces, and the precise nature of physical constants like the cosmological constant all indicate a fine-tuned universe. Scientists like Roger Penrose have calculated the odds of a universe emerging with the necessary conditions for life to be astronomically small, further supporting the argument that an underlying intelligence orchestrates these cosmic factors. This cosmic positioning reflects the Quranic verse: "And

We have placed within the heaven great stars and have beautified it for the observers" (Quran 15:16).

Similarly, the human body mirrors this divine architecture. The heart beats rhythmically, the circulatory system operates with mechanical efficiency, and our DNA encodes the blueprint of life with astonishing complexity. The double-helix structure of DNA follows precise mathematical ratios, reflecting an inherent order that scientists such as Francis Crick and James Watson have marveled at. The Fibonacci sequence appears in the branching of blood vessels and the arrangement of neurons, illustrating how divine order extends even within us.

The Quran acknowledges this miraculous design: "And in yourselves, do you not see?" (Quran 51:21). The Bible echoes this sentiment in Psalms 139:14: "I praise You, for I am fearfully and wonderfully made; Your works are wonderful, I know that full well." The Torah also reflects this theme in Job 10:11-12: "You clothed me with skin and flesh and knit me together with bones and sinews. You gave me life and showed me kindness."

Modern geneticists confirm that genetic coding operates like an advanced programming language, a discovery that has led scientists to remark, "DNA is like a computer program but far, far more advanced than any software ever created." Such findings affirm that the intricate systems within us are not the result of random processes but deliberate, mathematical design.

By studying these mathematical structures inherent in our very being, we gain insight into the Most High's architectural genius, seeing how all aspects of creation—from the grand scale of galaxies to the microscopic precision of DNA—are interconnected by divine mathematical truths. This intricate design reinforces the idea that mathematics is not merely a tool of human invention, but a divine language woven into the fabric of existence.

CHAPTER TWO
The Mathematics of Human Life and Consciousness

Mathematics governs not only the cosmos but also the very essence of human existence. Our biological processes, mental faculties, and even spiritual awareness align with mathematical principles that underscore the balance of life.

Just as the universe is governed by mathematical principles, so too is human life. The rhythms of our bodies—heartbeat, breathing, and neural oscillations—follow predictable patterns. The Golden Ratio appears in human anatomy, from the proportions of the face to the structure of DNA. This suggests that the divine design is not only external but also internal, written into the very code of our existence.

One striking example is the circadian rhythm, the natural cycle that regulates sleep, metabolism, and hormonal functions. Research suggests that optimal sleep occurs between 10 PM and 2 AM, aligning with the body's natural repair processes. This cycle reflects divine order, as Islam prescribes Qiyam-ul-Layl (night prayers) during the latter part of the night, a period where spiritual clarity is heightened.

The mathematical patterns of time also influence human life. Pregnancy, for instance, follows a predictable cycle of approximately 40 weeks (9 lunar months), mirroring the lunar phases. Many ancient cultures, including those referenced in the Bible and Quran, aligned their calendars with the moon, recognizing its rhythmic influence over life on Earth.

Furthermore, the human heartbeat follows a precise rhythm, pumping roughly 100,000 times per day in a complex yet harmonious pattern that sustains life. The Quran acknowledges this miraculous design: "And in yourselves, do you not see?" (Quran 51:21).

Mathematics also shapes human consciousness and moral structures. The concept of symmetry, which denotes balance and harmony, is deeply embedded in ethical principles. Justice, fairness, and reciprocity reflect symmetrical relationships where actions and consequences are balanced. This mirrors the Quranic teaching: "Is there any reward for good other than good?" (Quran 55:60), emphasizing the universal law of cause and effect.

Moreover, mathematical patterns govern the way we process information and make decisions. The brain's neural networks operate through complex algorithms, optimizing for efficiency and survival. Concepts like the Pareto Principle (80/20 rule), which states that 80% of effects come from 20% of causes, apply not only in economics but also in social dynamics, productivity, and even spiritual growth.

The Fibonacci sequence also appears in human cognition, influencing how we perceive aesthetics, music, and art. Patterns that follow Fibonacci ratios are often considered more pleasing to the eye, suggesting that our sense of beauty is hardwired to recognize mathematical harmony.

On a spiritual level, mathematics governs our acts of worship. The five daily prayers in Islam and the three daily prayers in Judaism establish a structured rhythm, synchronizing the believer's connection with the divine. These structured patterns in worship reinforce discipline, mindfulness, and inner peace, mirroring the order found throughout creation.

The divine order extends beyond physical and cognitive realms to the moral fabric of humanity. The Ten Commandments, the Five Pillars of Islam, and the 613 commandments in the Torah represent structured moral systems based on numerical organization. These principles provide a mathematical framework for ethical behavior, guiding individuals toward righteousness and social harmony.

Even in language and communication, mathematical principles prevail. Zipf's law, which predicts the frequency of word usage in languages, demonstrates how mathematical patterns govern linguistic structures. This law reveals that a small number of words are used frequently, while the majority are used less often, reflecting efficiency and balance in human communication.

Ultimately, the mathematical harmony in nature and human life is a testament to the Most High's wisdom. It reveals a universe where everything is connected, purposeful, and beautifully ordered. By understanding these patterns, we not only gain insight into the cosmos but also into ourselves, recognizing that we are integral parts of a divine equation that spans the infinite.

By recognizing these mathematical underpinnings in human life, we see how the Most High has infused creation with numerical harmony, guiding us toward both physical and spiritual well-being.

CHAPTER THREE
The Symphony of Universal Laws

The universe operates with extraordinary precision, governed by universal laws that reflect the wisdom and intentionality of the Most High. These laws are not random occurrences but mathematical principles that ensure balance, harmony, and continuity across all dimensions of existence. Just as a symphony requires every instrument to be in tune, the cosmos is orchestrated through laws that maintain its rhythm and structure. Each law, like a note in a grand composition, contributes to the intricate melody of existence. From the smallest particle to the vastness of galaxies, these laws offer a glimpse into the divine wisdom that sustains all of existence. Some scientists and spiritual seekers refer to these as the 12 Universal Laws, recognizing their presence in both the natural and metaphysical realms.

These laws include:
The Law of Attraction
The Law of Cause and Effect
The Law of Compensation
The Law of Correspondence
The Law of Divine Oneness
The Law of Gender
The Law of Inspired Action
The Law of Perpetual Transmutation of Energy
The Law of Polarity

The Law of Relativity
The Law of Rhythm
The Law of Vibration

In this chapter, we will explore all 12 and demonstrate how they celebrate the Most High through His amazing equations, offering insight into the intricate harmony within the universe.

The 12 Universal Laws and Their Scientific Counterparts

These 12 Universal Laws are an extension of divine principles, offering deeper insights into the cosmic order. Many of these laws also correspond to fundamental physical laws that govern the natural world, demonstrating the seamless integration of science and spirituality.

The Law of Attraction & The Law of Gravity

The Law of Attraction states that like attracts like. What we focus on expands, and our thoughts, emotions, and energy influence what we draw into our lives. "Call upon Me; I will respond to you" (Quran 40:60). This verse signifies the reciprocal nature of divine connection, where one's sincere prayers and intentions invoke a response from the Most High. Similarly, in the Bible, Matthew 7:7 states, "Ask, and it will be given to you; seek, and you will find; knock, and it will be opened to you." This highlights a universal truth—our focused energy and desires shape our reality, much like the gravitational pull that draws objects toward one another. The Torah also emphasizes this principle in Deuteronomy 4:29: "But if from there you seek the Lord your God, you will find Him if you seek Him with all your heart and with all your soul." These scriptural references affirm the power of attraction in both the spiritual and physical realms, illustrating how divine responsiveness and universal attraction operate in harmony.

Correspondingly, the Law of Gravity, articulated by Isaac Newton, describes the force of attraction between objects, ensuring stability and order throughout the cosmos. Newton's equation, $F = G(m1*m2)/r^2$, illustrates how objects are drawn together, just as our intentions and thoughts influence our reality. Gravity, as a fundamental force of nature, ensures cohesion in the universe, preventing celestial bodies from drifting into chaos. The Quran acknowledges this divine order in Surah Al-Mulk 67:3, "[He] who created seven heavens in layers. You do not see in the creation of the Most Merciful any inconsistency. So, return your vision to the sky, do you see any breaks?" Similarly, the Bible in Job 26:7 states, "He stretches out the north over empty space; He hangs the earth on nothing," highlighting the divine orchestration of gravitational forces. Scientists like Albert Einstein further expanded our understanding of gravity through General Relativity, demonstrating how massive objects bend space-time itself. This concept mirrors the spiritual truth that our intentions and energy shape the reality around us, just as gravity shapes the fabric of the cosmos. In Islam, the significance of intention is paramount, as reflected in the hadith of the Prophet Muhammad (peace be upon him): "Actions are judged by intentions, and every person will be rewarded according to their intention" (Sahih Bukhari, Hadith 1). This principle underscores not only our physical actions but also the sincerity of our inner state influences our external reality. Similarly, in science, gravitational forces, while invisible to the naked eye, fundamentally shape the structure of the universe—just as unseen intentions shape the course of human experience. Relatable examples include how a kind-hearted person attracts goodwill and support, or how a focused and determined individual creates success through perseverance, both demonstrating the interplay of intention and attraction within the cosmic framework.

The Law of Cause and Effect & Newton's Third Law

Every action has a reaction. This principle is foundational to both science and spirituality. "Whatever good you put forward for yourselves—you will find it with Allah" (Quran 2:110).

Newton's Third Law states: "For every action, there is an equal and opposite reaction." This mirrors spiritual teachings, such as "You will be recompensed for what you used to do" (Quran 45:22) and "A man reaps what he sows" (Galatians 6:7, Bible).

This law can be seen in everyday life. If a person consistently acts with kindness and generosity, they are more likely to receive kindness in return. Conversely, actions rooted in deceit or negativity often lead to negative consequences. This principle also plays out in relationships, career advancement, and personal development. The Prophet Muhammad (peace be upon him) emphasized this truth, stating, "Do good, and good will be done to you" (Musnad Ahmad). Similarly, science shows that cause and effect govern everything from the behavior of atoms to the movement of galaxies, affirming this universal law as both a physical and spiritual reality. In quantum mechanics, the concept of causality plays a crucial role in understanding how subatomic particles interact, influencing the very fabric of reality. Likewise, in astrophysics, gravitational interactions shape planetary orbits, star formations, and even black holes, demonstrating that every cosmic action produces a corresponding effect. This mirrors the interconnected principles found in faith, where moral actions yield divine recompense, and in daily life, where our choices ripple outward, influencing personal and collective outcomes. Whether in the scientific or spiritual realm, the immutable law of cause and effect underscores the intricate balance woven into existence by the Most High.

The Law of Compensation & The Conservation of Energy

The universe ensures balance and fairness in all aspects of existence. "And that man will have nothing except what he strives for" (Quran 53:39). This verse highlights the essential truth that effort and intention are directly tied to the outcomes we experience, emphasizing the Most High's justice in rewarding sincere striving.

The First Law of Thermodynamics states that energy cannot be created or destroyed, only transformed. This principle of conservation reflects the divine balance in life—every action is accounted for, and nothing is wasted in the grand design.

In nature, we see this law exemplified in ecosystems where energy is cycled seamlessly—plants convert sunlight into energy, which nourishes animals, whose remains eventually enrich the soil to sustain future growth. Similarly, in human life, our actions matand choices ripple through time, transforming into opportunities, lessons, or consequences. Just as energy cannot be lost in physics, neither can the intentions and efforts we put forth; they return to us in ways often beyond our immediate perception. The Quran states, "And do not forget your share of the world, and do good as Allah has done good to you" (Quran 28:77), reinforcing the balance and reciprocity found in both nature and divine law.

Furthermore, this principle is evident in economic and social systems. Efforts invested in personal growth, education, or charitable acts often yield rewards, even if delayed. In business, ethical practices attract trust and long-term prosperity, while dishonest dealings lead to eventual downfall. This mirrors the spiritual truth that honorable deeds are never wasted, as emphasized in the Bible: "Give, and it will be given to you. A good measure, pressed down, shaken together and running over, will be poured into your lap. For with the measure you use, it will be measured to you" (Luke 6:38).

Additionally, scientific research supports this principle through the concept of neuroplasticity—the brain's ability to reorganize itself based on effort and learning. Every time we engage in positive habits or disciplines, neural pathways strengthen, reinforcing the idea that compensation follows sincere effort. This affirms the interplay between divine justice, human perseverance, and the transformative power of energy in the universe. "And that man will have nothing except what he strives for" (Quran 53:39). This verse highlights the essential truth that effort and intention are directly tied to the outcomes we experience, emphasizing the Most High's justice in rewarding sincere striving.

The Law of Correspondence & The Fractal Nature of Reality

The Law of Correspondence expresses the principle "As above, so below; as within, so without." This law suggests that patterns repeat across different scales of reality, from microscopic atoms to galactic structures. It highlights the interconnectedness of all things, reinforcing the idea that divine order exists in both the macrocosm and microcosm.

This concept is deeply embedded in religious texts and philosophical teachings. The Torah emphasizes this divine symmetry: "For the Lord is great and greatly to be praised; He is to be feared above all gods" (Psalm 96:4). This verse underscores reverence due to the Most High, reflecting how His divine presence is echoed across all realms of existence. Similarly, the Quran states: "And We have certainly created man, and We know what his soul whispers to him, and We are closer to him than [his] jugular vein" (Quran 50:16). This profound verse signifies the intimate connection between the Creator and His creation, mirroring the microcosmic and macrocosmic reflections of divine order.

In physics, the concept of fractals and self-similarity exemplifies this principle, as structures in nature, like snowflakes, mountain ranges, river networks, and even biological systems such as blood vessels and neurons, display repeating patterns at different scales. The Mandelbrot set, one of the most famous fractals, demonstrates how a simple mathematical equation can produce infinitely complex and self-replicating patterns, much like the repeating symmetry we see in nature.

The human body itself is a testament to the Law of Correspondence. The Golden Ratio ($\varphi \approx 1.618$) is evident in the proportions of the human form, from the spacing of facial features to the length of limbs relative to the torso. The spiral patterns in DNA, the branching of neurons, and the rhythmic beating of the heart all mirror cosmic phenomena such as spiral galaxies, planetary orbits, and the expansion of the universe. This indicates that the same mathematical principles governing the cosmos are imprinted within us.

One particularly fascinating example is how the structure of the universe resembles neural networks in the human brain. Astrophysicists have discovered that the distribution of galaxies in the universe forms a web-like structure strikingly like synaptic connections in the brain. This further confirms that the patterns governing the largest cosmic structures also govern the smallest, reinforcing the Quranic assertion: "We will show them Our signs in the horizons and within themselves until it becomes clear to them that it is the truth" (Quran 41:53).

In spirituality and consciousness, this law also manifests in human behavior and perception. If a person maintains peace within themselves, they project peace into the world around them. This is why self-reflection, prayer, and meditation are essential in nearly every spiritual tradition—what exists within shapes what we manifest externally. The Bible echoes this in Proverbs 23:7: "For as he thinketh in his heart, so is he." Similarly, Islam teaches that one's internal state determines their actions and outcomes, as the Prophet Muhammad (peace be upon him) said: "There is a piece of flesh in the body; if it is sound, the whole body is sound, and if it is corrupt, the whole body is corrupt. Truly, it is the heart" (Sahih Bukhari, Hadith 52).

Moreover, in economics and social structures, the Law of Correspondence is seen in how financial systems, political dynamics, and cultural trends often mimic natural cycles of growth, decay, and renewal. Just as nature has seasons of planting, harvesting, and regeneration, societies experience periods of prosperity, decline, and rebirth. These cycles are a reminder that nothing exists in isolation—everything is part of a greater interconnected system.

In essence, the Law of Correspondence affirms that we are reflections of the grand design, and by studying nature, science, and spirituality, we uncover the deeper mathematical and divine patterns that structure reality. This law encourages us to align our inner world with higher principles so that our external world mirrors harmony, wisdom, and divine order.

The Law of Divine Oneness & Quantum Entanglement

The Law of Divine Oneness states that everything in existence is deeply interconnected. Every thought, action, and event influences everything else in ways that may not always be immediately perceptible. This principle aligns with the Quranic verse:
"And He is with you wherever you are" (Quran 57:4), which emphasizes the ever-present nature of the Most High, affirming that all things exist within His

divine knowledge and will. Similarly, the Bible conveys this oneness in Colossians 1:16-17: "For by Him all things were created, in heaven and on earth, visible and invisible... and in Him, all things hold together."

In quantum physics, this spiritual truth is echoed in the phenomenon of quantum entanglement—a mysterious property where two or more particles become instantaneously linked, regardless of the distance between them. If one particle is altered, the other responds immediately, even if separated by billions of light-years. This defies classical physics and suggests that all things are fundamentally interconnected at the most fundamental level of reality.

The Science of Oneness: Entanglement and the Universe

Einstein famously referred to quantum entanglement as "spooky action at a distance," yet modern experiments have confirmed that particles remain mysteriously interdependent beyond the limits of space and time. Scientists at institutions such as MIT and CERN have demonstrated that entangled particles can communicate instantly, suggesting that there is no true separation between any parts of the universe.

As alluded to earlier, when discussing the interconnectedness of natural systems in previous chapters, this principle of divine oneness is not merely theoretical; it actively governs our physical, spiritual, and social realities. If at the smallest scale of reality, particles are never truly independent, then what does that imply for humanity and the broader universe? It reinforces the idea that all living beings, all thoughts, and even all events are part of a greater unified field. This corresponds with the Islamic concept of Tawhid—the absolute oneness of God and His dominion over all creation. Nothing exists in isolation; everything is sustained by and connected to the Creator.

Divine Oneness Reflected in Nature and Humanity

This interconnectedness manifests in nature, human relationships, and social structures. Consider the ecosystem—a perfectly balanced network where the survival of one species depends on another. When one element of an ecosystem is disturbed, the ripple effects can be catastrophic, proving the deep interdependence of all life.

Similarly, human interactions are built on unseen yet powerful connections. The Quran states:
"O mankind, indeed We have created you from male and female and made you peoples and tribes that you may know one another" (Quran 49:13).
This verse highlights that, while we are distinct individuals, we are fundamentally one global family, each playing a role in the broader structure of human existence.

Psychologists have also explored the collective unconscious, a term coined by Carl Jung, which refers to the shared knowledge and archetypal patterns embedded within all human beings. This suggests that the human consciousness itself is interconnected, just as entangled particles mirror each other. Studies in neuroscience have also shown that emotions and thoughts can be contagious—a concept supported by quantum biology, where even biological systems display properties of entanglement at a molecular level.

Spiritual Teachings on Interconnectedness

The concept of Divine Oneness is a recurring theme in many faith traditions:

The Torah describes God's presence filling all existence:
"Do I not fill the heavens and the earth? Declares the Lord" (Jeremiah 23:24).

Hinduism speaks of the interconnected "Brahman," the universal spirit present in all things.

Buddhism teaches about "Dependent Origination," stating that nothing arises independently but exists through a web of cause and effect.

Even in everyday human experience, we witness this interconnectedness. A kind word can transform someone's entire day. A single act of generosity can spark a chain reaction of goodwill. The butterfly effect concept in chaos theory—suggests that minor changes in one part of the world can lead to massive consequences elsewhere.

Practical Applications: Living in Alignment with Divine Oneness

Understanding that everything is connected means we must approach life with greater consciousness and responsibility.

Our thoughts matter — The Law of Oneness implies that every thought contributes to the collective consciousness. Negative thinking, gossip, and envy create low vibrational energy that spreads, just as love, gratitude, and kindness uplift not only the individual but those around them.

Our actions ripple outward — Just as a single drop of water creates waves in a vast ocean, every decision we make has consequences beyond what we can immediately see. Islam teaches that even the smallest act of goodness is accounted for:
"So whoever does an atom's weight of good will see it, and whoever does an atom's weight of evil will see it" (Quran 99:7-8).

Unity through diversity — The diverse cultures, backgrounds, and perspectives among people are not meant to divide but to strengthen human unity. The Torah emphasizes this concept in Leviticus 19:18, stating:
"Love your neighbor as yourself."
By embracing diversity as a reflection of the divine order, we uphold the sacred unity of creation.

Both science and scripture reveal that the universe is an interconnected whole. The Law of Divine Oneness and quantum entanglement demonstrate that separation is merely an illusion; everything influences everything else in profound ways. This principle urges us to act mindfully, compassionately, and with purpose, knowing that our existence is woven into the great divine tapestry of life.

The Law of Gender & The Principle of Complementarity

The Law of Gender states that masculine and feminine energies exist in all creations. These energies are not solely about biological distinctions but extend into all aspects of existence, including nature, thought, emotion, and even the very fabric of the universe. "Glorified is He who created all pairs—from what the earth grows and from themselves and from that which they do not know" (Quran 36:36). This verse highlights the inherent duality within creation, reinforcing that balance between these forces is necessary for harmony and growth.

In physics, the Principle of Complementarity, particularly demonstrated through wave-particle duality, shows that matter can behave as both a particle and a wave, depending on how it is observed. This paradox reflects the coexistence of seemingly opposite qualities within the same entity, much like the balance between masculine and feminine energies. Similarly, in quantum mechanics, the presence of both properties suggests that duality is a fundamental component of reality, mirroring the spiritual truth that balance is essential to creation.

This principle is also evident in nature. In the plant kingdom, pollination requires both male and female components, reinforcing the necessity of complementarity for reproduction and continuation of life. In ecosystems, predator-prey relationships, cycles of destruction and renewal, and even the balance between oxygen and carbon dioxide exchanges between plants and animals reflect a finely tuned duality.

In human psychology and society, this law manifests through the balance of logical and emotional intelligence, action, and reflection, giving and receiving. Masculine energy often represents attributes such as assertiveness, strength, and structure, while feminine energy embodies intuition, nurturing, and fluidity. The healthiest individuals, relationships, and societies integrate both aspects, ensuring a comprehensive approach to decision-making and interpersonal connections.

Ancient wisdom traditions also recognize this principle. In Chinese philosophy, Yin and Yang symbolize the interdependence of opposites—one cannot exist without the other. The Bible echoes this concept in Genesis 1:27, "… male and female He created them," illustrating that divine creation inherently contains both masculine and feminine elements.

The concept of gender complementarity extends beyond physical distinctions to spiritual and intellectual realms. In Islamic thought, balance between masculine and feminine principles are seen as a path to personal and communal harmony. The

Prophet Muhammad (peace be upon him) demonstrated this balance in his leadership, embodying both strength and mercy, justice, and compassion.

Just as in physics, where wave-particle duality suggests that neither aspect is superior but rather necessary for a complete understanding of reality, the interplay of gender energies in life is not about dominance but about equilibrium. Societal structures that respect and cultivate both forces create environments where creativity, wisdom, and progress flourish.

In summary, the Law of Gender and the Principle of Complementarity remind us that all things exist in a balance of opposites. Whether in the cosmos, nature, human relationships, or spiritual traditions, this balance is not a limitation but a source of strength. The recognition and harmonious integration of these energies allow for the flourishing of life, growth, and enlightenment.

The Law of Inspired Action & The Principle of Motion

Manifestation requires action. Intentions alone are not enough; we must take steps toward our goals. While thoughts and prayers set the foundation, without action, they remain mere aspirations. Inspired action is movement aligned with divine guidance, ensuring that efforts are directed toward meaningful progress.

Newton's First Law of Motion states that an object remains at rest or in motion unless acted upon by an external force. This principle highlights the necessity of effort to initiate change—whether in the physical world or in our personal, professional, or spiritual lives. Just as a body at rest will remain stationary unless a force compels it forward, a person who desires success must take deliberate steps to achieve it.

Spiritual Perspective: Faith with Action

Islam, Christianity, and Judaism emphasize the necessity of coupling faith with action. The Quran states:

"And that man will have nothing except what he strives for." (Quran 53:39)

This verse affirms that effort is essential—one cannot simply desire success or righteousness; they must take steps toward it. The Prophet Muhammad (peace be upon him) also taught the importance of action, saying, "Tie your camel, then trust in Allah." (Tirmidhi) This hadith underscores that while faith in the Most High is crucial, responsible effort is also required.

Similarly, the Bible echoes this principle:
"Faith without works is dead." (James 2:26)
This passage emphasizes that belief alone is insufficient faith must be demonstrated through action. The Torah reinforces this in Deuteronomy 28:1-2, where obedience and effort lead to blessings:
"If you fully obey the Lord your God and carefully

follow all His commands… all these blessings will come on you and accompany you if you obey the Lord your God."

Thus, across all three scriptures, the message remains clear: action is a divine requirement for transformation and blessings.

Personal Development: Progress through Action

In personal growth, individuals often set intentions for change—whether it's improving health, acquiring knowledge, or developing discipline. However, without action, intentions remain unfulfilled. Consider someone desiring to lose weight or improve their fitness. They may set goals, visualize success, and affirm their intentions, but unless they engage in regular exercise and healthy eating, no change occurs.

This aligns with the principle that movement generates momentum. The first step—whether enrolling in a class, starting a new habit, or making a crucial decision—is often the most challenging. Yet, it is this action that propels individuals toward their goals. Just as a rocket requires an initial burst of energy to escape gravity, people must exert effort to overcome inertia in their lives.

Business & Success: Taking Initiative

Many entrepreneurs and professionals aspire for success but struggle with execution. The Law of Inspired Action teaches that while vision and planning are essential, consistent action is what turns ideas into reality.

For instance, a person may dream of starting a business. They might create a vision board, attend seminars, and visualize success. However, without taking practical steps—such as registering the business, developing a product, or reaching out to potential clients, the dream remains unrealized.

Business leaders like Madam CJ Walker and Daymond John exemplify this principle. They did not achieve success by merely dreaming; they took massive action, persisted through obstacles, and continuously adapted their strategies. Their journey demonstrates that movement, even in uncertainty, leads to breakthroughs.

Madam C.J. Walker, America's first self-made female millionaire, exemplifies the Law of Inspired Action through resilience, faith, and hard work. Born to formerly enslaved parents, she worked as a laundress before developing her own line of hair care products for Black women—a market that had been largely ignored.

Despite facing racial and gender barriers, she acted by selling her products door-to-door, training thousands of agents, and expanding her business nationally.

She didn't wait for the perfect conditions—she created opportunities through bold action and persistence.

Her success wasn't just personal, she became a philanthropist and activist, uplifting other Black entrepreneurs and funding scholarships and social causes.

Her life reflects Newton's First Law of Motion—by taking that first step, she built unstoppable momentum. As she famously said: "I got my start by giving myself a start."

Daymond John, the founder of FUBU and an investor on Shark Tank, started his clothing brand from his mother's house with no outside funding. Instead of waiting for investors or ideal circumstances, he acted on his vision:

He hand-sewed his early clothing designs and personally sold them in the streets of Queens.

He leveraged creative strategies, such as getting LL Cool J to wear his brand in a GAP commercial, which skyrocketed its visibility.

His persistence and action led to FUBU becoming a billion-dollar brand.

John often states that success requires movement—dreams alone won't build businesses, but consistent action will. His philosophy aligns with the Quranic principle: "And that man will have nothing except what he strives for" (Quran 53:39).

By embracing the Law of Inspired Action, we recognize that faith without effort remains stagnant, but faith coupled with movement transforms the world around us.

The Science of Action & Momentum

From a scientific perspective, Newton's First Law of Motion is a perfect metaphor for human behavior.

Psychologists have studied the Zeigarnik Effect, which states that once we start a task, our brains are wired to want to complete it. This explains why the hardest part of any endeavor is simply beginning.

A writer facing writer's block will often overcome it just by writing a few sentences.

A student struggling with procrastination can defeat inertia by simply opening a book and reading a paragraph.

A person seeking a healthier lifestyle can break through resistance by choosing a single healthy meal.

Each small step builds momentum, proving that action fuels further action.

Final Reflection

The Law of Inspired Action reminds us that faith and effort must work together. Whether in spirituality, personal growth, or business, success is not achieved through passive wishing, it requires purposeful movement, perseverance, and divine alignment.

"Indeed, Allah will not change the condition of a people until they change what is in themselves." (Quran 13:11)

This verse reaffirms that transformation begins with deliberate action, harmonizing divine guidance with human responsibility. Whether in the cosmos or in our lives, motion is required for progress.

The Law of Perpetual Transmutation of Energy & The Laws of Thermodynamics

Energy is always in motion, constantly shifting, evolving, and transforming. This universal principle signifies that nothing remains stagnant change is an inherent characteristic of existence. Everything we experience, whether thoughts, emotions, or physical actions, creates energy that must be redirected or transformed. "Indeed, with hardship comes ease" (Quran 94:6). This verse reflects the reality that difficult moments are not permanent, and that transformation is always underway, aligning with the scientific understanding that energy is never lost but merely changes form.

The Laws of Thermodynamics, which govern the flow of energy, emphasize transformation, growth, and renewal. The First Law of Thermodynamics states that energy cannot be created or destroyed, only transferred, or transformed. This aligns with spiritual teachings that nothing in creation is ever wasted—every experience, challenge, or effort contributes to a larger purpose. Similarly, the Quran reminds us of divine order in Surah Al-An'am:

"And with Him are the keys of the unseen; none knows them except Him. And He knows what is on the land and in the sea. Not a leaf falls but that He knows it." (Quran 6:59)

Just as energy transitions between states, life's events, and challenges shift, paving the way for growth and new opportunities. Likewise, the Second

Law of Thermodynamics, which speaks of entropy, states that systems tend toward disorder unless energy is purposefully introduced. This principle highlights the necessity of conscious effort in personal, social, and spiritual development. Without effort, focus, and intention, a person's spiritual, mental, and emotional state can deteriorate. The Torah reflects this reality in Proverbs 14:23:

"All hard work brings a profit, but mere talk leads only to poverty."

This verse underscores that just as energy requires direction to maintain order, so too do human efforts require action to create meaningful results.

Examples of Transformation in the Natural World

We see the Law of Perpetual Transmutation of Energy manifesting everywhere in nature and human experience. For example:

The Water Cycle: Water transforms through the process of evaporation, condensation, and precipitation. What begins as a droplet in the ocean can later return as rain to nourish the land, demonstrating the endless cycle of transformation.

Metabolism in the Human Body: Food energy is constantly converted into fuel, ensuring that life processes continue uninterrupted.

The Butterfly's Metamorphosis: A caterpillar does not remain in its original form; it undergoes a profound transformation inside the chrysalis before emerging as a butterfly.

The Death and Rebirth of Stars: In space, stars collapse into supernovae, creating the very elements that form new planetary systems.

These natural processes reflect the Most High's wisdom in designing a universe where energy and existence are never wasted but continually refined and redirected.

Personal Transformation and Spiritual Growth

This law also applies to human beings, emphasizing that people are not bound by their circumstances; growth is always possible. The key is understanding that effort, faith, and perseverance are the catalysts for transformation.

Turning Struggles into Strength: Hardships often become the foundation for personal strength. Just as raw iron is heated and shaped into a strong tool, human character is refined through adversity.

Faith as an Agent of Change: Many people find that sincere worship and prayer shift their reality, much like how focused energy can alter physical outcomes. The Bible reminds us in 2 Corinthians 4:16:

"Though outwardly we are wasting away, yet inwardly we are being renewed day by day."

The Power of Positive Thinking and Actions: Just as negative emotions can weigh down the soul, positive actions elevate one's spiritual and mental state. A single act of kindness, like a small ripple in a pond, spreads and transforms the environment around it.

The Prophet Muhammad (peace be upon him) said:

"The most beloved of deeds to Allah are those that are consistent, even if they are small." (Sahih Bukhari)

This Hadith reinforces the principle that small but consistent positive actions have the power to bring about profound transformation over time.

Harnessing Energy for Purposeful Change

One of the most practical applications of this law is learning to channel energy productively. Many people feel stuck in life because they do not actively direct their energy toward change. However, scientific, and spiritual wisdom alike teaches that transformation requires intention and movement.

Redirecting Negative Energy: Just as fire can either destroy or provide warmth, emotions like anger or disappointment can either consume us or be transformed into motivation for positive action.

Growth in Knowledge and Wisdom: The mind is like a muscle—it must be exercised through study, reflection, and experience. Energy invested in learning and self-improvement always leads to elevation in understanding and ability.

Social and Global Change: The greatest revolutions in history began with individuals who understood that their thoughts and actions had power. By harnessing their energy toward justice, equality, and truth, they transformed societies.

From the movement of celestial bodies to the metabolic processes in the human body, perpetual transmutation ensures that nothing remains idle. The Bible echoes this in 2 Corinthians 4:16:

"Though outwardly we are wasting away, yet inwardly we are being renewed day by day."

This renewal is a constant state, driven by both divine decree and natural law. Ultimately, this law calls upon us to recognize that transformation is inevitable—but whether that change is constructive or destructive depends on how we channel our energy.

Just as the sun converts nuclear energy into light and heat, individuals must transmute their challenges into wisdom, their pain into perseverance, and their efforts into tangible rewards. Through divine wisdom and scientific understanding, we see that every struggle, every effort, and every moment of life is part of an ever-unfolding cycle of transformation.

The Law of Polarity & The Expansion and The Contraction of the Universe

The Law of Polarity teaches that everything in existence has its opposite: light and darkness, joy and sorrow, heat and cold, life and death. This duality is not meant to create division but rather to establish harmony, contrast, and growth. Without darkness, light would not be appreciated. Without sorrow, joy would lack meaning. The interplay of opposites is what allows transformation to occur.

The Quran highlights this universal balance: "And of everything We have created pairs, that you may remember" (Quran 51:49). This verse serves as a reminder that polarity is not a coincidence, it is by divine design, embedded in every aspect of creation. The Bible reinforces this idea: "To everything there is a season, and a time for every purpose under heaven" (Ecclesiastes 3:1), showing that life's experiences are structured around cyclical transitions between opposing states. The Torah similarly recognizes the dual nature of existence in Isaiah 45:7: "I form the light and create darkness; I bring prosperity and create disaster; I, the Lord, do all these things." This passage acknowledges that both struggle and success come from the Most High, emphasizing that opposition is an inherent and purposeful part of creation.

Polarity in the Physical Universe: The Expansion and Contraction of the Cosmos

Polarity exists on a grand scale in the cosmos itself, particularly in the expansion and contraction of the universe. Since the Big Bang, scientists have observed that the universe has been expanding—a discovery credited to astronomer Edwin Hubble. However, some theories suggest that at some point, this expansion may reverse, leading to a Big Crunch, where the universe collapses inward and possibly begins anew in a cyclic process.

This concept of cosmic expansion and contraction mirrors the law of polarity: for every outward force, there is an inward pull. The Quran references this phenomenon in Surah Al-Anbiya 21:104: "The Day when We will fold the heaven like the folding of a [written] sheet. As We began the first creation, We will repeat it. [That is] a promise binding upon Us. Indeed, We will do it." This verse aligns with scientific theories of cyclical cosmology, suggesting that creation is not a one-time event but rather a continuous cycle of birth, expansion, contraction, and rebirth.

Even Einstein's Theory of General Relativity supports the possibility of a contracting universe, depending on the balance of gravitational forces. If the universe's density reaches a critical point, it could stop expanding and reverse direction—a process that mirrors the balance of expansion and contraction in all natural systems.

This pattern is also reflected in black holes, which are areas in space where matter collapses under intense gravity, forming singularities where expansion ceases entirely. Yet, some scientists theorize that black holes could give rise to new universes, reinforcing the idea that expansion and contraction are complementary, not contradictory, forces in the universe.

Polarity in Nature and Human Experience

Polarity is deeply embedded in nature. Some examples include:

The Earth's magnetic poles: The Earth maintains balance with its north and south magnetic poles, creating a stable electromagnetic field that protects life from solar radiation.

Electricity and charge: Positive and negative charges allow electricity to flow, making all modern technology possible. Without this duality, electrical current—and life as we know it—would not function.

Homeostasis in biology: The human body constantly balances opposing forces, for example, the sympathetic nervous system (which triggers "fight or flight") and the parasympathetic nervous system (which calms the body). This ensures stability and health.

Beyond the physical world, polarity shapes human emotions and experiences. Struggle and hardship cultivate resilience and wisdom, just as night gives way today. The Quran reassures believers: "Indeed, with hardship comes ease" (Quran 94:6), reinforcing the divine law that difficulties are always followed by relief.

Psychology also supports this idea. Post-traumatic growth is a well-documented phenomenon where individuals emerge from adversity stronger, wiser, and more compassionate. Just as opposing forces in physics create energy, challenges in life create personal evolution and transformation.

Embracing Polarity for Growth

The Law of Polarity teaches that contrast is essential for progress. Just as the universe oscillates between expansion and contraction, our lives move through cycles of growth, challenge, renewal, and balance.

By embracing polarity rather than resisting it, we can:

- Find purpose in adversity: Understanding that hardships lead to wisdom and strength.
- Develop patience and gratitude: Recognizing that ease follows difficulty, just as dawn follows night.
- Live in harmony with nature and divine order: Seeing balance in all aspects of life, from relationships to personal goals.

This profound truth guides us toward higher understanding, reminding us that opposites do not conflict—they complement, refine, and uplift one another.

The Law of Relativity & The Theory of Relativity

The Law of Relativity teaches that everything gains meaning through comparison, nothing exists in absolute isolation. Every experience, challenge, or blessing in life is only understood in relation to something else. This law urges gratitude, resilience, and personal growth, reminding us that our perception shapes reality.

The Quran affirms this principle in Surah Al-Inshirah 94:5-6, stating:

"For indeed, with hardship comes ease. Indeed, with hardship comes ease."

This repetition is significant; it serves as a reminder that struggles are always accompanied by relief, even when we do not immediately perceive it. Challenges are not permanent but rather steppingstones to growth and wisdom. The Bible reinforces this in Romans 8:18:

"I consider that our present sufferings are not worth comparing with the glory that will be revealed in us."

Similarly, the Torah teaches in Deuteronomy 8:2-3 that the hardships faced by the Israelites in the wilderness were meant to test and humble them, ultimately leading to spiritual refinement.

This law teaches us that our perception of hardship or success is shaped by what we compare it to. A person who has lived in poverty and later attains financial stability may feel deeply grateful, while another person born into wealth may take their fortune for granted. Perspective is everything.

The Theory of Relativity: Understanding Time and Space in a Relative Universe

In physical science, Albert Einstein's Theory of Relativity revolutionized our understanding of the universe, demonstrating that time and space are not fixed but relative to the observer.

There are two key aspects:

Special Relativity – The speed of light is constant, but time and space shift depending on motion. If a person moves near the speed of light, time slows down for them—this is known as time dilation.

General Relativity – Gravity bends space-time, meaning time moves slower in stronger gravitational fields. Astronauts in space age slightly faster than people on Earth due to weaker gravity.

This mirrors the spiritual reality of the Law of Relativity: just as time and space are influenced by perspective, so are our experiences of trials and blessings.

A person struggling financially in one country may still be considered wealthy compared to someone in another part of the world.

A challenge that seems unbearable today may, in hindsight, be viewed as a necessary lesson for personal and spiritual development.

Two individuals experiencing the same event may react differently—one might feel devastated, while the other sees it as an opportunity for growth.

The Law of Relativity teaches us to cultivate gratitude and adaptability, knowing that every situation is part of a greater journey. The Quran supports this idea in Surah Ibrahim 14:7:

"If you are grateful, I will surely increase you [in favor]."

This verse highlights that recognizing and appreciating blessings brings about more abundance, while focusing only on hardships blinds us to the ease that follows.

The Law of Relativity in Nature and Human Perception

This principle is woven into the very fabric of existence:

Temperature Sensation: Water that feels warm to one person may feel cold to another, depending on previous exposure.

Speed & Motion: A passenger on a plane moving at 600 mph feels still inside but is traveling at high-speed relative to the ground.

Cultural & Historical Perspectives: What is considered a challenge or hardship today may have been normal or even desirable in a different era or culture.

Biological Adaptation: The human body adjusts to different altitudes, temperatures, and conditions over time. A person used to extreme cold may find mild winter temperatures comfortable, while someone from a warmer climate may perceive them as unbearable.

In a broader sense, the Law of Relativity encourages us to practice patience, humility, and mindfulness. Just as scientists recognize that time and space are fluid and influenced by numerous factors, our own lives and experiences are shaped by our perspectives and what we choose to focus on.

The Spiritual and Scientific Significance of Perspective

The Law of Relativity challenges us to shift our focus. Instead of seeing obstacles as unfair punishments, we can view them as opportunities for refinement.

The Quran supports this mindset shift in Surah Al-Baqarah 2:216:

"But perhaps you hate a thing, and it is good for you; and perhaps you love a thing, and it is bad for you. And Allah knows, while you know not."

Similarly, the Bible in Ecclesiastes 3:1-4 speaks of the divinely appointed balance in life:

"To everything there is a season, a time for every purpose under heaven: a time to be born and a time to die, a time to plant and a time to uproot... a time to weep and a time to laugh."

The Torah also emphasizes this balance, reminding us that challenges and blessings are part of a greater, divine cycle. In Genesis 50:20, Joseph tells his brothers:

"You intended to harm me, but God intended it for good, to accomplish what is now being done, the saving of many lives."

Real-Life Applications of the Law of Relativity
Personal Growth & Overcoming Hardship

A student who fails an exam may initially feel devastated, but later realize it was a wake-up call to develop better study habits, leading to future academic success.

A job loss may seem like a setback but could open the door to better career opportunities.

Relationships & Emotional Intelligence

Understanding that others have different perspectives can reduce conflict and increase empathy.

What one person perceives as an insult, another might see as constructive criticism—it's all relative to past experiences and mindset.

Gratitude & Mindfulness

A person may feel discontent with their home until they visit a place where people have no shelter.

Someone struggling with finances may feel burdened, but in the eyes of someone with nothing, they may seem blessed.

These examples emphasize that our perception of hardship or success is shaped by what we compare it to. The Law of Relativity teaches patience and trust in divine wisdom, reminding us that life's trials are opportunities for personal and spiritual growth.

By embracing this law, we shift from a mindset of scarcity to abundance, from despair to hope, and ultimately recognize that all experiences serve a greater purpose in our journey of faith and enlightenment.

The Law of Rhythm & The Cyclical Nature of the Cosmos

The Law of Rhythm states that everything in existence moves in cycles, flowing between highs and lows, growth and decline, activity, and rest. Nothing remains static of creation is in perpetual motion, reflecting the natural rise and fall of energy throughout the universe. This law governs biological, cosmic, and spiritual rhythms, reminding us that harmony is found in embracing life's natural cycles rather than resisting them.

The Quran highlights this divine order in Surah Al-Furqan 25:62:

"And He it is Who has made the night and the day in succession for whoever desires to remember or desires gratitude."

This verse signifies that change and cycles are intentional and purposeful, providing humans with opportunities for reflection, renewal, and appreciation. The Bible similarly states in Ecclesiastes 3:1-2:

"To everything there is a season, and a time for every purpose under heaven: a time to be born and a time to die, a time to plant and a time to uproot."

This concept of divine rhythm is also emphasized in the Torah, particularly in the Sabbath cycle (Exodus 20:8-11), which designates a day of rest every seven days, aligning human activity with the natural cycle of work and renewal.

The Law of Rhythm in Physical Science & Nature

This law is deeply rooted in physics, astronomy, biology, and ecology, governing natural patterns such as:

The Motion of Celestial Bodies: Planets orbit stars in predictable cycles, governed by Kepler's Laws of Planetary Motion. The Earth rotates on its axis every 24 hours, producing the rhythm of day and night, while the moon orbits the Earth every 27.3 days, affecting ocean tides.

The Seasons & Solar Cycles: The Earth's axial tilt causes the four seasons, demonstrating a cyclical balance between growth and dormancy. Additionally, the 11-year solar cycle dictates the rise and fall of solar activity, influencing space weather.

The Biological Clocks of Life: The circadian rhythm regulates sleep-wake cycles in humans, ensuring synchronization with light and darkness. Similarly, menstrual cycles, migration patterns of birds, and hibernation in animals all follow precise rhythmic patterns.

In physics, wave mechanics also illustrates the Law of Rhythm. Sound, light, and electromagnetic waves move in oscillating patterns, demonstrating that all energy moves in rhythmic cycles. This principle is seen in the quantum realm, where subatomic particles vibrate in waves, emphasizing the inherent rhythm of the physical and spiritual universe.

The Spiritual & Psychological Impact of Rhythmic Cycles

Understanding the Law of Rhythm allows us to embrace change rather than resist it. In life, we experience seasons of prosperity and challenge, much like the changing tides of the ocean. Recognizing these cycles helps us cultivate patience and resilience.

Emotional & Mental Well-being

Just as waves rise and fall, so do our emotions. There are days of high energy, motivation, and productivity, followed by moments of exhaustion and introspection. This is natural and does not indicate failure but rather a time for inner growth.

The Quran reassures us in Surah Ash-Sharh 94:6: "Indeed, with hardship comes ease."
This verse reminds us that difficult moments are followed by relief, just as night transitions into day.

The Bible echoes this in Psalm 30:5: "Weeping may endure for a night, but joy comes in the morning."

The Rhythm of Faith & Worship

Daily prayers, fasting, and spiritual disciplines follow rhythmic patterns that align with cosmic and biological cycles.

Islamic prayer (Salah) occurs five times a day, aligning believers with the natural shifts in daylight and reinforcing spiritual discipline.

Ramadan follows the lunar cycle, allowing fasting to synchronize with the changing seasons over time.

Jewish observances, such as the Sabbath and the seven-year Shemitah cycle, emphasize rest and renewal, paralleling ecological rhythms.

Life's Cycles of Growth & Renewal
Economic markets rise and fall, reflecting the ebb and flow of wealth and opportunity.

Relationships have cycles of connection, challenge, and renewal, fostering deeper understanding and resilience.

Periods of hardship are followed by phases of abundance, reinforcing that no phase is permanent change is inevitable, and balance will always return.

Practical Applications: Aligning with the Law of Rhythm

Honor Rest & Reflection: Just as the Earth enters cycles of rest (winter), humans must also prioritize rejuvenation to sustain productivity and creativity.

Recognize & Adapt to Life's Waves: Rather than fear setbacks, understand that they are temporary phases leading to new growth.

Cultivate Resilience & Mindfulness: Knowing that hardship is followed by ease, we can approach challenges with faith and patience.

Follow Nature's Timing: Aligning activities, diets, and work schedules with natural cycles (such as sleeping when it's dark) enhance overall well-being and efficiency.

Final Reflection

The Law of Rhythm reminds us that life is a beautifully composed symphony, where every rise and fall, every period of struggle and ease, is divinely orchestrated. It teaches patience, trust, and adaptability, allowing us to flow with change rather than resist it.

The Quran beautifully encapsulates this truth in Surah Al-Lail 92:1-4:

"By the night as it covers, and by the day as it appears in brightness, and by He who created the male and the female—Indeed, your efforts are diverse."

This verse emphasizes the balance of opposites and the rhythmic cycles that define existence. Similarly, the Bible reminds us in Galatians 6:9:

"Let us not grow weary in doing good, for at the proper time we will reap a harvest if we do not give up."

By embracing this divine rhythm, we harmonize with the natural flow of the universe, understanding that every challenge is temporary, every difficulty is followed by ease, and every season serves a purpose in our journey of faith and enlightenment.

The Law of Vibration & The Quantum Field

The Law of Vibration states that everything in the universe is in constant motion, vibrating at specific frequencies. This universal principle suggests that energy flows through all things, influencing our thoughts, emotions, relationships, and the physical world around us. It asserts that nothing is truly at rest; even seemingly solid objects are made up of atoms and subatomic particles in perpetual motion. At its core, this law highlights that everything—our thoughts, emotions, and even physical matter—operates on distinct vibrational frequencies that influence the reality we experience.

Scriptural Foundations of Vibration and Energy

The Quran touches on this concept of movement and energy:

"And the mountains will pass away with an awful passing away" (Quran 52:10), alluding to the impermanence and constant motion in creation. This echoes the scientific understanding that even massive celestial bodies are in constant motion, subject to the gravitational pull and energy fluctuations of the universe.

Similarly, the Bible states in **Psalm 104:5**, *"He set the earth on its foundations; it can never be moved."* Though this verse speaks of divine stability, it also hints at the Most High's design of a universe in perfect, dynamic balance—one where everything moves, shifts, and vibrates in its ordained manner.

The Torah also references movement and divine orchestration in **Ecclesiastes 3:1**:
"To everything, there is a season, and a time to every purpose under the heaven."
This verse encapsulates the idea that the Most High has created the universe with an inherent rhythm and vibration that guides all existence.

Scientific Counterpart: The Quantum Field & Vibrational Energy

Modern physics aligns with this law through **Quantum Field Theory (QFT)**, which suggests that all matter and energy arise from fluctuating quantum fields. Particles are simply excitations in these fields, meaning that everything we perceive as "solid" is fundamentally composed of vibrating energy. **Albert Einstein** famously stated:
"Everything in life is vibration."

Moreover, **String Theory**, one of the leading models in theoretical physics, proposes that the smallest building blocks of the universe are vibrating strings of energy rather than point-like particles. These vibrations determine the characteristics of particles, much like different frequencies in music create different notes.

Wave-Particle Duality further supports this principle. Physicists have discovered that electrons and photons can behave as both particles and waves depending on how they are observed. This reinforces the idea that all matters exist in a vibrational state and can shift based on external forces, much like how spiritual and emotional states fluctuate with experiences.

Spiritual Insight:
The Quran acknowledges this duality of perception and existence in **Surah Al-Hadid 57:20**:
"Know that the life of this world is but amusement and diversion and adornment and boasting to one another and competition in increase of wealth and children—like the example of rain whose growth pleases the disbelievers. Then it

dries, and you see it turn yellow; then it becomes scattered debris."

This verse metaphorically describes the impermanence of material things, paralleling how quantum particles shift between states of being depending on perception and influence.

Vibrations in Daily Life: How Frequencies Affect Us

Vibrational frequencies shape our emotional and physical well-being in profound ways.

The Impact of Sound and Healing Frequencies

Sound waves are one of the most direct ways we experience vibration. Different sound frequencies impact our mood, cognition, and even physical health.

1. **432 Hz**: Often called the *natural frequency of the universe*, known for its calming and harmonizing effects.

2. **528 Hz**: Sometimes referred to as the *"miracle tone,"* believed to promote healing and DNA repair.

3. **7.83 Hz**: Known as the *Schumann Resonance*, which aligns with the Earth's electromagnetic field, enhancing mental clarity and spiritual awareness.

Spiritual Insight:

The Quran emphasizes the power of sound in creation:

"And when your Lord said to the angels, 'Indeed, I will create a human being out of clay from an altered black mud. And when I have proportioned him and breathed into him of My

soul, then fall down to him in prostration.'" (Quran 15:28-29)

Here, the Most High's command ("Be") is a vibrational force that initiates creation, much like how frequencies shape physical matter in the natural world.

The Bible also speaks of the power of sound in **Genesis 1:3**:
"And God said, 'Let there be light,' and there was light."
Again, this suggests that sound waves—or vibrations—played a fundamental role in creation.

Cymatics: Visualizing Sound & Vibration

Cymatics is the study of visible sound vibration. When sound frequencies pass through a medium like water or sand, intricate geometric patterns emerge. This suggests that vibrational energy shapes reality in structured ways.

Spiritual Insight:

The Quran describes how divine harmony shapes creation:
"The sun and the moon move by precise calculation, and the stars and the trees prostrate. And the heaven He raised and imposed the balance." (Quran 55:5-7)

This aligns with cymatics, where precise vibrational forces create structured, geometric forms.

Human Vibrational Fields & Emotional Frequencies

Scientific studies show that human emotions emit measurable electromagnetic frequencies. Higher emotions like love, gratitude, and joy emit high frequencies, while emotions like anger, fear, and resentment operate at lower frequencies.

The Quran addresses vibrational energy shifts through consciousness and the remembrance of Allah, a practice known as dhikr.
"Those who believe and whose hearts find comfort in the remembrance of Allah—surely in the remembrance of Allah do hearts find comfort." (Quran 13:28)

Dhikr serves as a powerful tool for realigning one's spiritual state, raising vibrational energy, and maintaining a deep connection with the Most High. By engaging in consistent remembrance—whether through reciting Subhanallah (Glory be to Allah), Alhamdulillah (All praise is due to Allah), Allahu Akbar (Allah is the Greatest), or La ilaha illa Allah (There is no god but Allah)—one reinforces a state of inner tranquility, divine mindfulness, and heightened spiritual frequency.

Scientific studies on sound frequencies and neuroplasticity suggest that repetitive affirmations and meditative recitations can calm the nervous system, lower stress levels, and promote emotional resilience. Just as sound waves influence the physical world—creating geometric patterns in cymatics—the rhythmic recitation of dhikr influences the human heart, mind, and soul, harmonizing them with the divine order of the universe.

The Prophet Muhammad (peace be upon him) emphasized the transformative power of dhikr,

stating:
"The example of the one who remembers his Lord and the one who does not is like that of the living and the dead." (Sahih Bukhari, Hadith 6407)

This highlights the life-giving energy that remembrance of Allah brings to the soul, aligning it with a higher spiritual frequency, much like how tuning an instrument ensures harmony within an orchestra. Just as the universe vibrates in divine precision, dhikr attunes the believer to the rhythm of the Most High's creation, strengthening faith, emotional well-being, and spiritual insight.

The Bible echoes this in **Proverbs 17:22**: *"A joyful heart is good medicine, but a crushed spirit dries up the bones."*

These scriptures affirm that elevating one's vibrational state through prayer, faith, and positive emotions leads to greater peace, clarity, and healing.

Harnessing the Power of Vibrational Energy

Understanding the Law of Vibration empowers us to take intentional action in raising our energy and consciousness. This can be achieved through:

Spiritual Practices:

1. **Prayer & Meditation**: Enhances spiritual awareness and elevates frequency.
2. **Reciting Sacred Texts**: The Quran and other scriptures contain sound frequencies that uplift consciousness.

3. **Gratitude & Positive Speech:** The words we use carry energy; speaking blessings invite divine alignment.

Scientific Applications:
1. **Listening to Healing Frequencies:** 432 Hz and 528 Hz promote relaxation and well-being.
2. **Practicing Mindfulness:** Staying present reduces stress and raises vibrational energy.
3. **Exposure to Nature:** Trees, water, and sunlight emit natural frequencies that rejuvenate our energy field.

Final Thought: The Law of Vibration as a Path to Divine Alignment

Understanding the Law of Vibration allows us to consciously shape our experiences and recognize that everything around us is interconnected through energy. Whether through scientific discoveries in quantum mechanics, sound healing, or spiritual teachings, this law reveals that the Most High's creation operates through a perfect vibrational order.

By embracing this truth, we deepen our connection with the divine, aligning our thoughts, emotions, and actions with the harmonious rhythm of existence.

CHAPTER FOUR
Equations as Reflections of the Divine Mind

Wave Function & Observer Effect: The Role of Consciousness in Reality

Formation Modern quantum mechanics has revealed a startling truth: the act of observation itself influences reality. This phenomenon, known as wave function collapse, suggests that particles exist in a state of potentiality until they are measured or observed, at which point they assume a definite state. This principle, derived from the famous double-slit experiment, demonstrates that matter and energy behave as both particles and waves—existing in superposition until consciousness interacts with them.

The Observer Effect on Quantum Mechanics

In the double-slit experiment, light and electrons were observed behaving as waves when not measured, meaning they existed in multiple possible locations simultaneously. However, when measured by an observer, the wave function "collapsed," forcing the particle to take a single, definite position. This implies that reality does not take on a fixed form until it is observed.

Physicist John Wheeler proposed that we live in a participatory universe, where observation is essential for reality to take shape. This supports the idea that consciousness and perception are fundamental to the way the universe operates.

Yet, this concept is not new, it has long been reflected in divine scriptures.

Faith and the Observer Effect: Consciousness as a Divine Tool

The Quran, Bible, and Torah emphasize that faith, prayer, and divine intention shape the unseen realm, mirroring the way observation shapes reality at the quantum level.

Quran (Al-Baqarah 2:186):
"And when My servants ask you concerning Me, indeed I am near. I respond to the invocation of the supplicant when he calls upon Me."

This verse suggests a dynamic interaction between divine reality and human consciousness—prayer, like observation, brings the unseen into manifestation.

Bible (Matthew 21:22):
"And whatever you ask in prayer, you will receive, if you have faith."

Here, faith functions as a spiritual observer, directing energy toward intended outcomes, much like the conscious act of measurement in quantum mechanics collapses a wave function into a defined reality.

Torah (Proverbs 23:7):
"As a man thinketh in his heart, so is he."

This verse reinforces the idea that thought itself contributes to shaping reality—a concept deeply aligned with quantum consciousness theories.

The Quran also confirms this in Surah Al-An'am (6:59):

"And with Him are the keys of the unseen; none knows them except Him. And He knows what is on the land and in the sea. Not a leaf falls but that He knows it."

This passage highlights the divine role in unseen realities, mirroring the quantum state of potentiality that collapses into reality upon observation.

Does This Mean Consciousness is the Key to Reality?

Many leading physicists, including Max Planck (the father of quantum theory), believed that consciousness is primary to the universe, not secondary.

Planck once stated:
"I regard consciousness as fundamental. I regard matter as derivative from consciousness. We cannot get behind consciousness. Everything that we talk about, everything that we regard as existing, postulates consciousness."

This aligns deeply with faith-based perspectives, which hold that God's knowledge and observation are what sustain reality itself.

Quran (Al-Imran 3:29):
"Say, 'Whether you conceal what is in your breasts or reveal it, Allah knows it. And He knows what is in the heavens and what is on the earth. And Allah is over all things competent.'"

This verse describes a divine observer effect—everything is already known by the Most High, and nothing can escape His knowledge. Just as quantum mechanics suggests that observation collapses reality, faith teaches that divine observation sustains creation itself.

Scientific Method for Spirituality: A Bridge Between Faith and Empirical Study

Traditionally, science and spirituality have been seen as separate disciplines, but recent advancements in neuroscience, quantum physics, and consciousness studies suggest that spiritual phenomena can be tested through structured scientific inquiry. If spiritual principles such as prayer, meditation, and divine consciousness influence the material world, then they should be measurable in ways similar to natural laws.

Can Faith Be Measured?

While faith is often considered an internal experience, research has shown that spiritual practices affect brain function, physiology, and even physical health. Some key areas of scientific inquiry include:

1. Prayer and Healing Studies

Studies by organizations such as the National Institutes of Health (NIH) and Duke University have explored how intercessory prayer affects patient recovery.

Results suggest that patients who are prayed for, even without their knowledge, experience statistically significant improvements in health outcomes compared to those who are not.

2. Meditation and Brain Function

EEG and fMRI studies show that meditation and prayer activate the prefrontal cortex and limbic system, reducing stress and enhancing cognitive function.

Research from Harvard Medical School has demonstrated that long-term meditators experience increased gray matter density, showing that spiritual practice physically alters brain structure.

3. Water and Consciousness Studies (Dr. Masaru Emoto)

Emoto's famous experiments involved exposing water molecules to spoken words, prayers, and intentions, then freezing them.

The molecular structures of the water formed intricate, beautiful patterns when exposed to positive affirmations and sacred texts, while negative words created distorted, chaotic structures.

This suggests that vibrational energy from thoughts and prayers impacts physical reality, reinforcing the Law of Vibration.

Testing Spiritual Laws Scientifically

If spiritual principles hold empirical weight, they can be tested using a structured scientific framework similar to how physical laws are studied. This could involve:

Controlled Studies:

Examining the effects of prayer and intention on physical and psychological well-being.

Measuring vibrational frequency changes in individuals before and after spiritual practices.

Exploring the effects of consciousness on material reality using quantum experiments.

Neuroscientific Approaches:

fMRI scans showing brain activity during prayer or divine contemplation.

Heart rate variability studies measure the physiological impact of spiritual devotion.

Scientific Studies on the Healing Power of Prayer

While faith-based healing has been practiced for millennia, modern science has sought to measure the physiological and psychological effects of prayer. Various controlled studies and meta-analyses have provided compelling evidence that prayer has tangible effects on health, stress reduction, and recovery rates.

1. The Harvard & Duke Studies on Prayer and Healing

Harvard Medical School and Duke University have conducted extensive research on how prayer affects healing outcomes.

A Harvard study found that patients who regularly engaged in prayer or meditation experienced significantly lower stress levels, faster recovery times, and enhanced immune function.

Duke University Medical Center discovered that patients undergoing cardiac surgery who were prayed for had fewer complications and a faster recovery compared to those who were not prayed for.

Another Duke study found that patients who engaged in daily prayer had lower levels of cortisol (the stress hormone), lower blood pressure, and a reduced risk of heart disease.

Quran (Surah Al-Ra'd 13:28):
"Indeed, in the remembrance of Allah do hearts find rest."

This aligns with findings that prayer and spiritual remembrance reduce stress, anxiety, and heart-related illnesses.

2. The "Mantra Prayer" Study – fMRI & Brain Activity in Prayer

A study conducted at the University of Pennsylvania used functional MRI (fMRI) to analyze brain activity during prayer.

Participants who engaged in deep prayer or repetitive recitations (such as Islamic dhikr, Christian rosary prayers, or Buddhist mantras) showed increased activity in the prefrontal cortex and limbic system, regions associated with focus, peace, and emotional regulation.

Neuroplasticity studies suggest that repeated prayer strengthens neural pathways, enhancing mental clarity, resilience, and emotional stability.

Bible (Philippians 4:6-7):
"Do not be anxious about anything, but in every situation, by prayer and petition, with thanksgiving, present your requests to God. And the peace of God, which transcends all understanding, will guard your hearts and your minds in Christ Jesus."

This biblical passage reinforces what fMRI studies show—prayer has a calming effect on the brain, reducing stress and activating the neurological centers for peace.

3. The NIH Study on Intercessory Prayer & Recovery from Illness

The National Institutes of Health (NIH) conducted a double-blind study on the effects of intercessory prayer (prayer done by others for a patient) on hospital patients.

Patients in the prayer group showed improved recovery rates, reduced need for pain medication, and lower incidence of post-surgical complications.

Those who received no prayer intervention had slightly longer recovery periods and reported higher levels of discomfort.

Torah (Exodus 15:26):
"For I am the Lord, who heals you."
This verse reflects the spiritual belief that divine intervention through prayer can bring about physical healing, aligning with the measurable effects seen in medical studies.

Scientific Evidence on Vibrational Healing & Prayer

Beyond traditional medicine, scientific research on vibrational energy supports the idea that prayer, meditation, and intention affect biological frequencies and overall health.

1. The HeartMath Institute – Prayer & Heart Coherence

HeartMath Institute researchers studied how prayer affects heart rhythms and coherence.

When people engage in prayer, gratitude, and divine remembrance, their heart rate variability (HRV) improves, leading to better immune function, lower stress, and increased overall well-being.

Electrocardiogram (ECG) scans show that deep spiritual states create synchronized, harmonic waves in the heart, promoting physical and mental healing.

Quran (Surah Al-Shu'ara 26:80):
"And when I am ill, it is He who cures me."
This verse aligns with the scientific findings that prayer-induced physiological changes contribute to recovery and well-being.

2. Masaru Emoto's Water Experiment – The Power of Words & Prayer

Dr. Masaru Emoto conducted famous experiments where he exposed water molecules to several types of words, prayers, and music, then froze the water to observe its crystal formations.

Positive prayers and words ("love," "gratitude," and scripture readings) created beautiful, symmetrical crystalline structures.

Negative words ("hate," "anger") produced chaotic, fragmented, and distorted structures.

Water exposed to Quranic recitations and Christian prayers formed highly structured, geometric formations, suggesting that spiritual energy influences physical matter.

Bible (John 1:1):
"In the beginning was the Word, and the Word was with God, and the Word was God."
This passage hints at the creative power of divine words, aligning with findings that prayers and affirmations alter molecular structures.

Final Thoughts: Prayer as a Scientific and Spiritual Healing Tool

The power of prayer is no longer just a spiritual belief; it is scientifically measurable. From brain scans showing neurological benefits to medical studies confirming improved health outcomes, the evidence is overwhelming:

1. Prayer reduces stress and promotes emotional well-being.

2. Spiritual practices improve heart health, lower blood pressure, and boost immune function.

3. Intercessory prayer correlates with faster healing and recovery in hospital patients.

4. Vibrational energy studies show that prayer influences biological frequencies and even molecular structure.

Quran (Surah Fussilat 41:53):
"We will show them Our signs in the horizons and within themselves until it becomes clear to them that this is the truth."

This verse reflects the ongoing scientific discoveries that align with divine truths, proving that faith and science are not separate but deeply interconnected.

Similarly, the Bible (Proverbs 25:2) states:
"It is the glory of God to conceal a matter; to search out a matter is the glory of kings."
This implies that seeking knowledge—including knowledge of spiritual truths—is a Divine endeavor.

The Torah (Hosea 4:6) reinforces this pursuit:
"My people are destroyed for lack of knowledge."
It is through knowledge, study, and experience that humanity comes to a greater understanding of the Divine.

CHAPTER FIVE
The Hidden Dimensions of Reality

Introduction: Expanding Perception Beyond the Visible

We are conditioned to believe that what we see, hear, and touch defines reality. But what if everything we experience is just a **thin veil over a far greater existence?** The **universe is not limited to what we perceive, it is layered, multidimensional, and alive with unseen forces.** Modern physics, ancient scriptures, and metaphysical insights all point to one undeniable truth: **our reality is only a fragment of the whole.**

Science has now begun to confirm what spiritual traditions have known for millennia—**that existence extends beyond the physical world.** The question is: **What are these hidden dimensions? How do they function? And most importantly, how do they impact our lives?**

In this chapter, we'll explore:

1. **Higher-Dimensional Existence:** The scientific and spiritual evidence for extra dimensions.
2. **Spacetime as a Data Structure:** Why reality functions more like a computational framework than solid matter.
3. **Conscious Agents & Spiritual Beings:** How higher-dimensional entities interact with our reality.

4. Parallel Universes & the Unseen Realms:
The intersection of science and scripture regarding multiversal existence.

This is not speculation. **This is the next frontier of human understanding.**

Higher-Dimensional Existence: The Fabric of Reality is Not What We Think

Physicists have long struggled to unify the laws of nature into a single framework. The deeper we probe, the more we realize that **our three-dimensional space, moving through time, is insufficient to explain the forces that govern reality. Math itself demands additional dimensions.**

1. **String Theory & Extra Dimensions**

If String Theory is correct, **we don't live in a four-dimensional universe (three space + one time). We exist in at least ten dimensions.** Six of them are compactified—hidden from us—yet fundamental to existence. This means that **forces, consciousness, and even spiritual energies may operate in dimensions beyond our perception.**

2. **Kaluza-Klein Theory & Unified Fields**

Decades before String Theory, Kaluza and Klein proposed that **gravity and electromagnetism could be unified if a fifth dimension existed.** If true, what we perceive as **spiritual energy may not be separate from physical forces—it may be part of the structure of reality itself.**

3. Quantum Superposition & Multidimensional States

Quantum mechanics has revealed that particles exist in **multiple states simultaneously until observed**. This suggests that reality itself is in flux until it is interacted with—a concept eerily like how spiritual experiences are described in **ancient texts**.

Spiritual Parallel:
The Quran describes **multiple layers of existence**:
"And We have certainly created above you seven layered heavens, and never have We been of the unaware." (Quran 23:17)

Similarly, Paul in **2 Corinthians 12:2** writes:
"I was caught up to the third heaven—whether in the body or out of the body I do not know—God knows."

If **higher dimensions exist**, then what we call "the spiritual world" is simply a deeper layer of reality.

Spacetime as a Data Structure: Reality is Not Solid—It's Information

We have been trained to think of the universe as **a collection of physical objects floating in empty space**. But what if **space itself is not fundamental? What if reality is structured more like a computational system than a physical one?**

1. John Wheeler's "It from Bit" Hypothesis

Wheeler, a pioneering physicist, argued that **the foundation of reality is not matter, but**

information. In other words, the universe operates **like a vast quantum computation, processing bits of data rather than chunks of physical substance.**

2. The Holographic Universe & Information Encoding

Studies in quantum gravity suggest that all the information in the universe may be encoded on a **two-dimensional boundary—like a hologram projecting a three-dimensional world.**

3. Consciousness as a Reality Processor

What if **the material world exists only when it is observed?** Modern physics suggests that **consciousness collapses probabilities into reality**—an idea that aligns with scripture's description of divine creation through awareness and will.

Spiritual Parallel:

The Quran states:

"And with Him are the keys of the unseen; none knows them except Him. And He knows what is on the land and in the sea. Not a leaf falls but that He knows it." (Quran 6:59)

In **Psalm 33:9**, the Torah echoes this idea:
"He spoke, and it came to be; He commanded, and it stood firm."

What if what we call "reality" is **actively sustained by divine consciousness?**

Conscious Agents & Spiritual Beings: The Intelligent Network of the Unseen

If reality is information-driven, then what role do **spiritual entities** play? I suggest that **higher-dimensional beings interact with us, not as abstract spirits, but as conscious agents within a structured, intelligent system.**

1. **Donald Hoffman's Consciousness Model**

Neuroscientist Donald Hoffman proposes that **our perceptions are merely an interface—a simplified model for interacting with a larger, unseen consciousness network.**

2. **The Role of Vibrational Frequencies**

In every religious tradition, spiritual beings communicate through **energy rather than speech.** Angelic encounters, prophetic visions, and divine guidance **are often described in terms of light, sound, and vibration.**

3. **Religious Descriptions of Unseen Beings**
    1. The **Quran** speaks of **angels, jinn, and unseen forces** that influence human affairs.
    2. The **Bible** describes **guardian angels and divine messengers** that interact with humans.
    3. The **Torah** recounts numerous encounters with divine presences in visions and dreams.

Scientific Parallel:

Quantum entanglement proves that **particles remain interconnected across vast distances, reacting** instantaneously. This mirrors the idea that **spiritual forces exist outside our spacetime constraints yet still affect us.**

Parallel Universes & the Unseen Realms: Science & Spirituality Converge

We are no longer asking *if* other realities exist. We are now asking *how many*.

1. **Multiverse Theories in Physics**
   1. The **Many-Worlds Interpretation** of quantum mechanics suggests that **every decision splits reality into parallel versions.**
   2. Some **cosmologists propose that universes exist side by side**, separated only by subtle differences in physics.
2. **Religious Descriptions of Multiple Realms**
   1. The **Quran** refers to **seven heavens**, indicating multiple layers of existence.
   2. The **Bible** describes different heavenly realms, **suggesting a tiered reality.**
   3. The **Torah** outlines a divine structure where **spiritual and material realms coexist.**
3. **Scientific Evidence of Possible Multiversal Interactions**

Some physicists believe that **gravitational**

anomalies or quantum fluctuations might provide proof of interactions between universes.

Spiritual Implication:
If multiple realities exist, then **divine intervention, prophetic visions, and supernatural experiences** may be interactions between these realms.

Final Thoughts: The Interwoven Nature of Reality

My insights and innovative physics **converge on one profound truth:**

1. **Reality extends beyond what we perceive.**
2. **Consciousness actively shapes existence.**
3. **Higher-dimensional beings are part of an intelligent, structured order.**
4. **The universe functions more like a data system than a mechanical structure.**

Spiritual Reflection:
"We will show them Our signs in the horizons and within themselves until it becomes clear to them that it is the truth." (Quran 41:53)

By exploring **the hidden dimensions of reality,** we uncover **the mathematical, scientific, and spiritual architecture of creation**—one that is alive with **purpose, intelligence, and divine harmony.**

CHAPTER SIX
THE FABRIC OF THE UNSEEN

Introduction: Beyond the Limits of Perception

Our senses deceive us. We assume that what we can see, hear, and touch defines reality. But modern physics, ancient scripture, and countless metaphysical experiences tell a different story. Reality is not confined to what is directly observable, it extends into the unseen, into forces, dimensions, and energies beyond our comprehension.

Science now confirms what spiritual traditions have taught for millennia: an invisible fabric underlies all existence, shaping our world in ways we are only beginning to understand. This chapter explores the nature of this unseen fabric, how it influences the known universe, and why understanding it is key to unlocking both scientific and spiritual truths.

Key topics include:

Dark Matter & Dark Energy: The unseen forces that shape the cosmos.

The Interplay of Energy & Consciousness: How vibrational frequencies structure reality.

The Role of the Unseen in Human Experience: The scientific and spiritual dimensions of intuition, inspiration, and divine guidance.

At the heart of this exploration is a radical but necessary shift in perspective: the unseen is not imaginary—it is the foundation upon which all things rest.

Dark Matter & Dark Energy: The Invisible Architecture of the Cosmos

The Scientific Mystery of the Unseen Universe

In the 20th century, physicists discovered a perplexing truth: most of the universe is invisible. Observations of galaxies revealed that their motions could not be explained by visible matter alone—there had to be an unseen force holding them together. This led to the discovery of dark matter, a mysterious substance that interacts with gravity but emits no light or energy.

Even more astonishing was the discovery of dark energy, a force that drives the accelerating expansion of the universe. These two unseen components make up a staggering 95% of the cosmos, meaning that what we perceive—stars, planets, galaxies—accounts for only 5% of reality.

The Quran describes the unseen nature of creation:
"And of everything We have created pairs, that you may remember." (Quran 51:49)

Could it be that dark matter and dark energy are the unseen counterpart to the visible universe—an integral, yet hidden, aspect of divine design?

Spiritual Parallels: The Hidden Forces in Scripture

Dark matter and dark energy provide a modern scientific analogy for spiritual realities described in ancient texts. Just as dark matter exerts gravitational influence without being directly observable, spiritual forces act upon the world in ways beyond our direct perception.

As mentioned in an earlier chapter, the Bible affirms this hidden structure of reality:
'For we wrestle not against flesh and blood, but against principalities, against powers, against the rulers of the darkness of this world, against spiritual wickedness in high places.'
(Ephesians 6:12)

Similarly, the Torah describes unseen angelic and divine forces operating within creation:
"Then he opened their eyes, and they saw the horses and chariots of fire all around Elisha." (2 Kings 6:17)

These passages hint at a fundamental truth realm beyond our senses interacts with the physical world, shaping events, guiding consciousness, and maintaining cosmic order.

The Interplay of Energy & Consciousness: How Vibrational Frequencies Shape Reality

Scientific Perspectives on the Unseen Energy Field

Modern physics has revealed that everything in existence vibrates at specific frequencies. The energy field that underlies reality is not static, it is dynamic, shifting, and responsive. Quantum field theory suggests that all particles arise from underlying fields of potentiality, vibrating into existence when disturbed by external forces.

Key discoveries supporting this concept include:

The Zero-Point Field: Quantum physics suggests that even in a vacuum, space is not empty, it is filled with fluctuations of energy, an invisible field from which all matter emerges.

Resonance & Harmonics: In physics, objects with similar frequencies can synchronize and amplify each other's energy, suggesting that consciousness itself may be a resonant force, influencing reality.

Nonlocality & Quantum Entanglement: Particles can be instantaneously connected across vast distances, implying that consciousness, intention, and prayer may affect the fabric of reality in ways beyond classical understanding.

The Quran describes this interconnected nature of existence:

"It is Allah Who created the heavens and the earth and whatever is between them in six days, and then He established Himself above the Throne. You have not besides Him any protector or intercessor; so, will you not be reminded?" (Quran 32:4)

This verse reflects the idea that the universe is sustained not merely by physical laws, but by divine orchestration—an energy beyond space and time, yet present in every moment.

The Role of the Unseen in Human Experience

If unseen forces shape the cosmos, do they also shape our personal realities?

Throughout history, spiritual traditions have spoken of intuition, divine inspiration, and supernatural guidance—phenomena that modern science is beginning to study through fields like quantum consciousness and neurotheology.

Intuition & Higher Knowledge: Accessing the Unseen Field

Many breakthroughs in science, philosophy, and creativity have come not through linear reasoning, but through sudden insights, intuitive knowing, and inspired moments of revelation. From Nikola Tesla's visions of alternating current to Einstein's discovery of relativity, many of history's greatest discoveries emerged not through mechanical calculation, but through connection with a deeper field of awareness.

The Quran speaks of divine knowledge given through unseen means:
"And He taught Adam the names—all of them. Then He showed them to the angels and said, 'Inform Me of the names of these, if you are truthful.'" (Quran 2:31)

This suggests that knowledge is not only gained through experience but can also be received from beyond—the unseen realm transmitting wisdom to those attuned to it.

Bridging the Scientific & Spiritual Understanding of the Unseen

Science and spirituality are converging toward a shared realization: reality is not solid, but energetic, structured, and deeply interconnected.

Dark matter & dark energy mirror spiritual forces described in scripture.

Quantum fields suggest a vibratory structure to reality, aligning with ancient teachings on divine resonance.

Consciousness interacts with these fields, meaning our thoughts, prayers, and intentions shape the very fabric of existence through the will and design of the Most High, who governs all unseen and seen forces.

The deeper we explore, the clearer it becomes: the unseen is not separate from us—it is the foundation of everything we experience. The ancients knew this long before modern physics caught up.

The Quran affirms this mystery:
"And you do not throw when you throw, but it is Allah who threw." (Quran 8:17)

This verse suggests that what we perceive as individual action is in fact guided, shaped, and amplified by unseen forces.

Final Thoughts: Learning to Perceive the Unseen

If reality extends beyond the visible, then our ability to perceive and interact with it is crucial. The ancients understood this, developing disciplines of prayer, meditation, fasting, and sacred geometry to align themselves with these deeper truths.

Spiritual practices like dhikr (remembrance), fasting, and prayer may function as methods for tuning into the unseen.

Science is beginning to validate what mystics and prophets have always known—that the material world is only part of the equation.

The next step in human evolution is not just technological advancement but a deeper understanding of the fabric of reality itself.

The Quran leaves us with a profound reminder: "And We have certainly created above you seven layered heavens, and never have We been of the unaware." (Quran 23:17)

To unlock the future, we must first acknowledge the hidden. The universe is speaking—through physics, through scripture, through direct experience. The question is, are we listening?

CHAPTER SEVEN
The Equation of Balance—Harmony in the Cosmos and Within

The Most High's design of the universe is not a collection of random occurrences but a carefully calibrated system of interdependent forces, all governed by equilibrium. Balance is an equation—one that manifests throughout nature, human existence, and even in the unseen dimensions. Whether we are speaking about the stability of celestial mechanics, the balance of ecosystems, or the moral and spiritual harmony embedded in divine law, the universe operates under a grand symmetry.

Mathematics confirms this through equations that regulate motion, energy, and order. The principle of equilibrium ensures that opposing forces never spiral into chaos but instead work in perfect alignment. The Quran reinforces this concept:
"And We have set the balance so that you may not transgress in the balance." (Quran 55:7-8)

This **balance** extends beyond physics—it governs human consciousness, ethics, and even the unseen realms. If balance is disrupted in any part of creation, whether in the physical, spiritual, or moral plane, disorder follows. The Most High's design does not allow for absolute chaos; instead, imbalance triggers forces that naturally work toward restoration.

The Equation of Balance in the Cosmos

Science affirms that the laws governing the cosmos adhere to precise balance. **The fine-tuning of the universe** is one of the strongest indicators of divine order. The gravitational pull of celestial bodies, the ratio of matter to antimatter, the nuclear forces holding atoms together—all must exist within razor-thin margins for stability to persist. Even a minor deviation in fundamental constants, such as the gravitational constant (G) or the cosmological constant (Λ), would result in a universe inhospitable to life.

Stephen Hawking acknowledged this delicate precision:

"If the rate of expansion one second after the Big Bang had been smaller by even one part in a hundred thousand million million, the universe would have collapsed before it ever reached its present size."

This is not coincidence; it is calculation. Every mathematical model describing the fabric of reality is governed by a **balancing equation** that ensures energy is neither lost nor wasted but remains in flux, transferring between states to maintain harmony.

The Bible confirms this truth in Proverbs 16:11: *"Honest scales and balances belong to the Lord; all the weights in the bag are of His making."*

Even in the quantum realm, **Heisenberg's Uncertainty Principle** dictates that precision in one measurement leads to uncertainty in another, demonstrating a built-in balance at the most fundamental level of reality. If we extend this principle beyond physics, it suggests that what appears as randomness or uncertainty is, in fact, a **divinely encoded structure** ensuring that everything remains in check.

The Equation of Balance in Nature

This divine equation is equally visible in Earth's ecosystems. The biosphere is an intricate, self-regulating system where every living organism plays a role in sustaining equilibrium. The nitrogen cycle, the oxygen-carbon dioxide exchange, the predator-prey relationship—all reflect the precision with which creation is governed.

Take, for example, the balance of **photosynthesis and respiration**. Plants absorb carbon dioxide and release oxygen, while humans and animals do the reverse, maintaining a cycle that allows life to flourish. This is not an accidental arrangement, it is a synchronized biological equation, proof of the Most High's intentionality in every aspect of existence.

The Quran affirms this in Surah Al-Hijr 15:19: *"And the earth—We have spread it and set therein firm mountains and caused to grow therein of every well-balanced thing."*

This verse suggests that nature itself has been calibrated with an internal balance, a design that supports sustainability. Science reinforces this through the concept of **homeostasis**, the body's ability to regulate itself through dynamic balance—whether through temperature regulation, pH levels, or blood sugar equilibrium.

When this balance is disrupted, disorder ensues. Excessive deforestation disrupts the oxygen cycle. Overhunting leads to ecosystem collapse. Pollution alters natural equilibrium. **Human actions impact the balance the Most High has ordained, creating consequences that ripple through creation.**

The Torah reinforces this responsibility in Leviticus 25:23-24:

"The land must not be sold permanently, because the land is Mine, and you reside in My land as foreigners and strangers. Throughout the land that you hold as a possession, you must provide for the redemption of the land."

This verse implies that balance in nature is not just physical but a **moral responsibility**. Humanity is entrusted with maintaining harmony within creation, acting as stewards rather than disruptors.

The Equation of Balance Within Ourselves

Just as the universe and nature require equilibrium, so too does the **human soul**. When we are out of balance—spiritually, mentally, or emotionally—we experience distress. The Quran teaches the importance of balance within the self:

"And do not make your hand as chained to your neck or extend it completely and thereby become blamed and insolvent." (Quran 17:29)

This verse metaphorically addresses **moderation,** instructing believers not to be excessive or negligent, but to maintain a **balanced approach** in all aspects of life—whether in spending, relationships, or spirituality.

The **heart, mind, and body** operate under the same principle. The nervous system has two main branches:

1. **The sympathetic nervous system (fight-or-flight),** which activates in the face of stress.
2. **The parasympathetic nervous system (rest-and-digest),** which restores calm.

Both must be in harmony. Excessive stress leads to anxiety and burnout, while too much passivity leads to stagnation. The key to well-being is **balance** between action and reflection, work and rest, devotion, and worldly responsibility.

The Prophet Muhammad (peace be upon him) emphasized this principle:
"Your body has a right over you, your eyes have a right over you, and your wife has a right over you." (Sahih Bukhari)

Similarly, the Bible states in Ecclesiastes 3:1:
"To everything there is a season, and a time for every purpose under heaven."

Spirituality itself requires balance. Too much focus on the material world leads to detachment from divine purpose, while excessive withdrawal from society neglects the duty to serve. The Most High does not require asceticism—He requires equilibrium.

The Mathematical Blueprint of Balance

Equilibrium is encoded into the fundamental equations of existence:

1. **Newton's Third Law:** Every action has an equal and opposite reaction.
2. **The Law of Conservation of Energy:** Energy is neither created nor destroyed, only transferred.
3. **The Fibonacci Sequence:** The ratios in nature reflect mathematical harmony.

In the **spiritual realm**, balance is found in:

1. **Divine Justice:** The Most High rewards sincerity and punishes wrongdoing, ensuring fairness.
2. **Moral Equilibrium:** Ethical laws in scripture preserve social stability.
3. **Consciousness & Choice:** Free will allows us to maintain or disrupt our own inner harmony.

The Torah highlights this responsibility in Deuteronomy 30:19:

"I have set before you life and death, blessings, and curses. Now choose life, so that you and your children may live."

This verse reinforces that balance is not imposed—it is chosen. The Most High provides the framework, but maintaining equilibrium is up to us.

Final Reflection: Living in Harmony with the Equation of Balance

Understanding the Equation of Balance allows us to align with **universal order**. Whether through scientific laws, natural systems, or personal well-being, balance is the governing principle of existence. Those who cultivate **moderation, justice, and alignment with divine law** experience **peace and stability**—while those who violate it experience **chaos and unrest.**

"Indeed, Allah commands justice, good conduct, and giving to relatives and forbids immorality, bad conduct, and oppression." (Quran 16:90)

This is the divine formula: **harmony in the cosmos, harmony in nature, and harmony within.** When we embrace balance as a principle, we walk in accordance with the Most High's will, not in resistance to divine order but as reflections of it.

CHAPTER EIGHT
The Equation of Time—Cycles, Seasons, and Divine Order

Time is one of the most elusive and profound aspects of existence. It governs the movement of celestial bodies, the progression of life, and the unfolding of events within the physical and metaphysical realms. Yet, despite its apparent linearity, time is far more complex than it seems. In physics, time is relative, shaped by gravitational forces and velocity, bending under extreme conditions. In spirituality, time is sacred, woven into divine decrees and eternal truths.

Time is not an arbitrary construct—it is a divinely structured equation, set in motion by the Most High to maintain order in creation. The Quran declares: "And He is the One who created the night and the day, and the sun and the moon; each is floating in an orbit." (Quran 21:33)

Similarly, Ecclesiastes 3:1-2 states:
"To everything, there is a season, a time for every purpose under heaven: A time to be born, and a time to die; a time to plant, and a time to uproot."

These scriptures reveal an inherent rhythm in time—a cycle ordained by divine intelligence, shaping existence according to mathematical precision. Whether it is the orbits of planets, the shifting of seasons, or the phases of human life, time is the fundamental equation that underlies all reality.

The Equation of Time in Physics: Relativity and the Fabric of Spacetime

Modern physics has uncovered the elastic nature of time, demonstrating that it is not fixed but dependent on gravitational forces and motion. Albert Einstein's Theory of Relativity revolutionized our understanding of time by showing that:

Time slows down in stronger gravitational fields.

Time moves faster for objects in weaker gravitational fields.

The faster an object moves; the slower time passes relative to a stationary observer.

This means that time is not an absolute constant—it is influenced by energy, mass, and space itself. Scientists have confirmed this by placing atomic clocks on airplanes and satellites, proving that time moves differently depending on position and motion.

The Quran alludes to this concept in Surah Al-Ma'arij 70:4:
"The angels and the Spirit will ascend to Him during a Day the extent of which is fifty thousand years."

This verse suggests time dilation, a phenomenon where time is experienced differently depending on the realm of existence. Similarly, in Psalm 90:4, the Bible states:
"For a thousand years in Your sight are but as yesterday when it is past, and as a watch in the night."

These verses align with scientific discoveries, affirming that time is relatively fluid, rather than rigid, adapting to divine law.

The Divine Order of Time in Nature

Time manifests through natural cycles, structuring the fabric of life. From the orbits of celestial bodies to biological rhythms, everything adheres to a preordained sequence that sustains existence.

The Earth's Rotation creates the day-night cycle, regulating biological functions and energy distribution.

The Moon's Phases determine lunar cycles, influencing ocean tides and biological patterns.

Seasonal Cycles govern agriculture, migration, and reproduction, reinforcing the interconnectedness of all living things.

The Quran acknowledges these natural rhythms: "And We have made the night and the day two signs, and We erased the sign of the night and made the sign of the day visible so that you may seek bounty from your Lord and know the count of years and calculation." (Quran 17:12)

This verse highlights that time is not only a measurement but a guiding system, an equation by which creation moves toward its divine purpose.

The Torah reinforces this in Genesis 1:14: "And God said, 'Let there be lights in the firmament of the heavens to divide the day from the night; and let them be for signs and seasons, and for days and years.'"

This passage confirms that time was structured from the beginning, embedded into creation as a divine blueprint for order and continuity.

The Equation of Time in Human Experience

Just as time governs celestial and natural cycles, it also shapes human existence. Life follows a mathematical sequence:

Birth → Growth → Maturity → Decline → Death

Learning → Experience → Wisdom → Legacy

Opportunity → Effort → Outcome → Reflection

This sequence aligns with the Quranic principle of cause and effect:
"And that man will have nothing except what he strives for, and that his effort will be seen. Then he will be recompensed for it with the fullest recompense." (Quran 53:39-41)

Time is the measure by which our efforts are judged, and every moment is either utilized for growth or lost to neglect.

The Prophet Muhammad (peace be upon him) emphasized this in a well-known hadith:
"Take advantage of five before five: your youth before your old age, your health before your sickness, your wealth before your poverty, your free time before your preoccupation, and your life before your death." (Narrated by Al-Hakim)

This hadith underscores that time is not merely passing—it is a currency, an equation that determines reward or regret, based on how it is invested.

Similarly, the Bible warns against wasting time in Ephesians 5:16:

"Redeeming the time, because the days are evil."

This verse teaches that time is not neutral—it must be actively utilized for good, or it becomes a lost opportunity.

The Interplay Between Divine Time and Human Free Will

One of the most intriguing aspects of time is the paradox of divine predestination and human free will. If the Most High exists outside of time, and all events are foreknown, how do human choices still matter?

This question is resolved through the equation of divine decree:

The Most High has set the framework of time, governing its flow and ultimate outcomes.

Humans exist within this framework, given the agency to act within time's constraints.

Time itself acts as a testing ground, allowing choices to unfold while remaining subject to divine will.

The Quran addresses this in Surah Al-Qamar 54:49:

"Indeed, all things We created with predestination."

Yet, this does not negate free will, as Surah Al-Ra'd 13:11 explains:

"Indeed, Allah will not change the condition of a people until they change what is in themselves."

In other words, time is both fixed and fluid—its structure is determined, but human choices shape its unfolding. This aligns with Einstein's theory that time is not absolute but relative, molded by external conditions.

The Torah echoes this in Deuteronomy 30:19: "I have set before you life and death, blessing and cursing. Therefore, choose life that both you and your descendants may live."

This verse affirms that while divine laws establish time's framework, human beings are given the power of choice within that equation.

Eternal Time vs. Temporal Time

The final mystery of time is its eternal vs. temporal nature.

Temporal time is the linear sequence we experience—days, years, history, and progression.

Eternal time exists beyond the limits of past, present, and future—unchanging, infinite, and transcendent.

The Quran describes divine time in Surah Al-Insan 76:1:
"Has there [not] come upon man a period of time when he was not a thing [even] mentioned?"

Before human awareness, time existed, and after human existence, time will continue in dimensions beyond our comprehension.

The Bible reflects this in Revelation 22:13: "I am the Alpha and the Omega, the beginning and the end, the first and the last."

This confirms that time itself is a creation, a mechanism set into motion to serve divine purpose. Once its function is complete, it will merge into eternity.

Final Reflection: Aligning with the Equation of Time

Time is not merely passing, it is calculated, moving creation toward an ultimate fulfillment. Whether in the expansion of the universe, the cycles of nature, or human experiences, time is the Most High's equation for progress.

The wise embrace this truth, understanding that every moment carries weight, and that alignment with divine order ensures that time is utilized rather than wasted. As the Prophet Muhammad (peace be upon him) said:
"The feet of a servant will not move on the Day of Judgment until he is asked about his life and how he spent it." (Tirmidhi)

In the equation of time, every second is an opportunity—one that either advances us toward divine harmony or leaves us in regret. The choice is ours.

CHAPTER NINE
Conclusion & Reflection

In all that I've shared within these pages—from equations to vibrations, from wave functions to the eternal stillness of higher dimensions, I have sought not to preach but to present a framework. A way of seeing. A map. One that bridges the rigor of science with the reverence of spirituality.

What I have come to understand is this: reality is not passive. It is not simply a place we inhabit. It is a responsive field—malleable, intelligent, and divine. Consciousness is not a byproduct of matter. Consciousness is primary. It shapes matter, time, and energy. It is the light behind the curtain, the voice within the silence, the algorithm beneath the stars.

Each equation I've explored is not just symbolic logic. It is a fingerprint of the Most High—mathematical evidence of sacred intention. These patterns do not emerge randomly. They are seeded. They are planted. They are calling us to wake up and remember that we are not separate from the Divine purpose—we are created to live in alignment with it.

I believe that science, when truly pursued without ego, points toward God. And I believe that faith, when truly practiced without fear, makes room for inquiry. When these two—science and faith—are no longer seen as enemies but as companions, something extraordinary happens. We begin to touch the face of truth. Not abstract truth but lived truth. Breathing truth. Truth that hums in our cells and sings in our spirit.

Throughout this journey, we've considered the role of vibration, the mystery of the unseen, the elegant balance that underlies existence, and the possibility that time itself is a projection from higher consciousness. These are not merely philosophical ideas, they are invitations. Invitations to perceive differently. To awaken to the sacred data encoded in reality. To recognize the intelligent architecture beneath the veil of form.

What if meditation isn't merely a practice of silence, but a return to source code? What if love itself is a frequency divine algorithm harmonizing creation? What if what we call miracles are simply glimpses of a higher law, momentarily revealing itself through cracks in the lower matrix?

We are not alone in this work. Across dimensions, across times, our ancestors, angels, and guides are also conscious agents. They operate through frequencies we've only just begun to understand. And yet, even in our current form, we are capable of resonance. Of alignment. Of communion.

The ancients spoke in symbols. The prophets spoke in parables. The scientists speak in numbers. I see them all as threads in the same sacred tapestry. My goal has been to trace those threads—not to claim final answers, but to remind us that we are designed to question, to seek, to elevate.

You, dear reader, are not here by accident. The same divine intelligence that coded the spiral of DNA and the expansion of the cosmos also whispered your name into existence. You are part of the equation. You are a conscious being formed by the Creator's will—able to perceive His signs and respond in alignment.

And so I leave you with this: Do not merely believe. Observe. Do not merely worship. Wonder. Do not merely accept. Ask. The Most High is not diminished by your curiosity—He is glorified by it. Keep seeking. Keep solving. Keep listening to the silence between numbers.

Because in that space, you may hear the voice of God.

CLOSING BENEDICTION

O Most High, Divine Author of Light and Law—
As we close these pages, we return to You with gratitude.
We have reasoned, reflected, and reached for truths beyond sight.

May what is written serve as a beacon for those searching for harmony between science and spirit.
May the wisdom gleaned ignite deeper reverence for Your eternal presence and perfect order.

To every heart that receives these words—grant clarity, peace, and renewed connection to You.
May this be not an ending, but a sacred continuation of seeking.

All praise is due to You, the Knower of All Things.
Ameen.

AFTERWORD

In the Name of God, the Most Gracious, the Most Merciful.

As I reflect on the work presented in Celebrating the Existence of the Most High through Equations, I am reminded that the sacred path of knowledge is one we walk with both heart and mind. This book is a modern contribution to a timeless tradition—one that recognizes the harmony between divine revelation and rational inquiry.

Throughout Islamic history, many luminaries have demonstrated that faith and science are not adversaries, but allies in our pursuit of truth. One such figure is Imam Jaʿfar al-Sadiq, may Allah be pleased with him—a towering figure and spiritual guide whose teachings laid the intellectual groundwork for both theology and natural philosophy. His reflections on creation, the nature of the soul, and the layered structure of the universe continue to inspire believers across generations.

What this book accomplishes is a rekindling of that tradition. It invites us to think deeply, not just about numbers and formulas, but about the sacred architecture of existence itself. It reminds us that behind the physical world lies a divine order—subtle, intentional, and majestic.

I commend the effort to present these ideas in a way that honors the legacy of our faith traditions while engaging with contemporary thought. Whether one approaches from the perspective of Islam, Christianity, or Judaism, there is something here that speaks to the universal yearning to understand our place in the cosmos.

May this work encourage future seekers to explore with reverence, to question with humility, and to remember always that true knowledge brings us closer to the Creator.

Sheikh Ahmad Hamza Abdullah
Islamic Scholar & Community Leader
Los Angeles, CA

APPENDIX: VISUAL DIAGRAMS & REFLECTIONS

This section contains visual representations that complement the key themes of the book. Each diagram corresponds with concepts explored in specific chapters, offering deeper insights into how mathematics, consciousness, and divine design intersect.

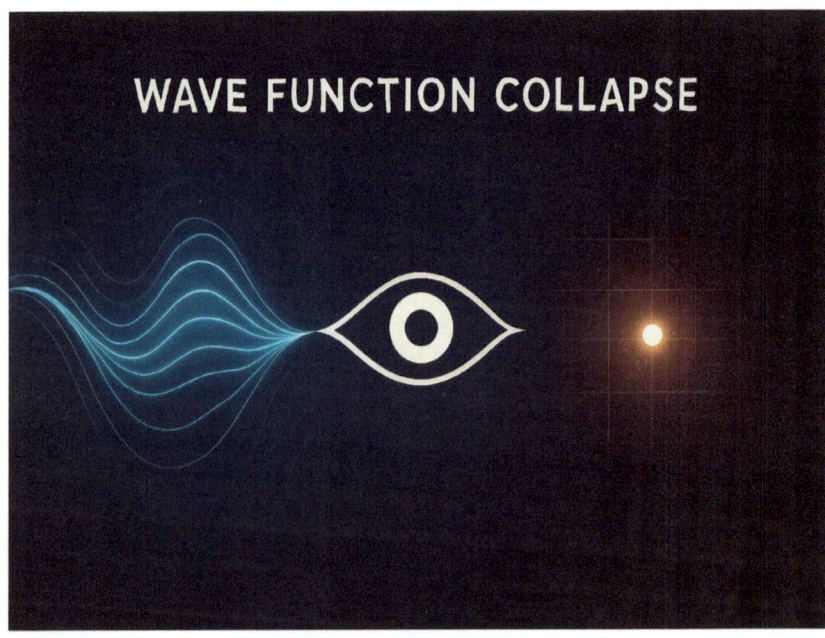

Diagram A1: The Wave Function and the Act of Observation

Associated Chapter: Chapter Four – *Equations as Reflections of the Divine Mind*

Diagram Description:

This digital illustration visualizes the central concept of wave function collapse—a foundational idea in quantum physics and my book's metaphysical framework. The scene is split into two distinct layers representing potential and manifest reality.

1. **Upper Layer: The Quantum Field**

This portion of the image is ethereal, filled with undulating waves, glowing particles, and fluid lines. These patterns represent wave functions—infinite probabilities existing in a superposition state. The waves shimmer in soft gradients of indigo, violet, and gold, symbolizing the unmanifest possibilities of creation. The motion is calm but chaotic, with energy constantly in flux.

2. **Observer Node (Center Focus)**

At the heart of the image is a radiant figure or point of light—symbolizing consciousness or the observer. This figure emits a subtle energy beam that connects the upper, quantum potential layer to the lower, physical reality layer. This visual metaphor reflects Tyrone's teaching that observation is the trigger that collapses potential into reality.

3. **Lower Layer: Manifest Reality**

As the observation occurs, a focused beam funnels

one wave from above into a crystallized form below. This lower section contains structured, geometric shapes—like cubes, spheres, and fractal grids—signifying matter, form, and spacetime. It's drawn with sharper edges, increased contrast, and solid colors to emphasize tangibility. This contrast with the fluid upper layer reinforces the transition from potential to actual.

4. **Light Bridges and Information Flow**
Arcs of light travel back and forth between the layers, suggesting that observation not only collapses reality but feeds information *back* into the quantum field. This represents the reciprocal nature of consciousness and creation—core to Tyrone's beliefs.

5. **Color Palette and Vibration**
The upper realm pulses with cooler tones (purples, silvers, electric blues), symbolizing divine intelligence and possibility. The lower world uses more earthy or structured tones (gold, teal, deep violet), grounding the image in realized experience. This duality supports Tyrone's theme of vibrational resonance shaping existence.

Narration: This illustration visualizes the collapse of the wave function as described in quantum physics. Prior to observation, particles exist in a state of probability—a "cloud of potential." Once observed, the wave collapses into a defined reality. This quantum behavior mirrors the spiritual principle that **conscious attention** and **intention** shape reality. It reflects how the divine mind, whether through prayer, awareness, or conscious choice—plays an active role in the unfolding of creation.

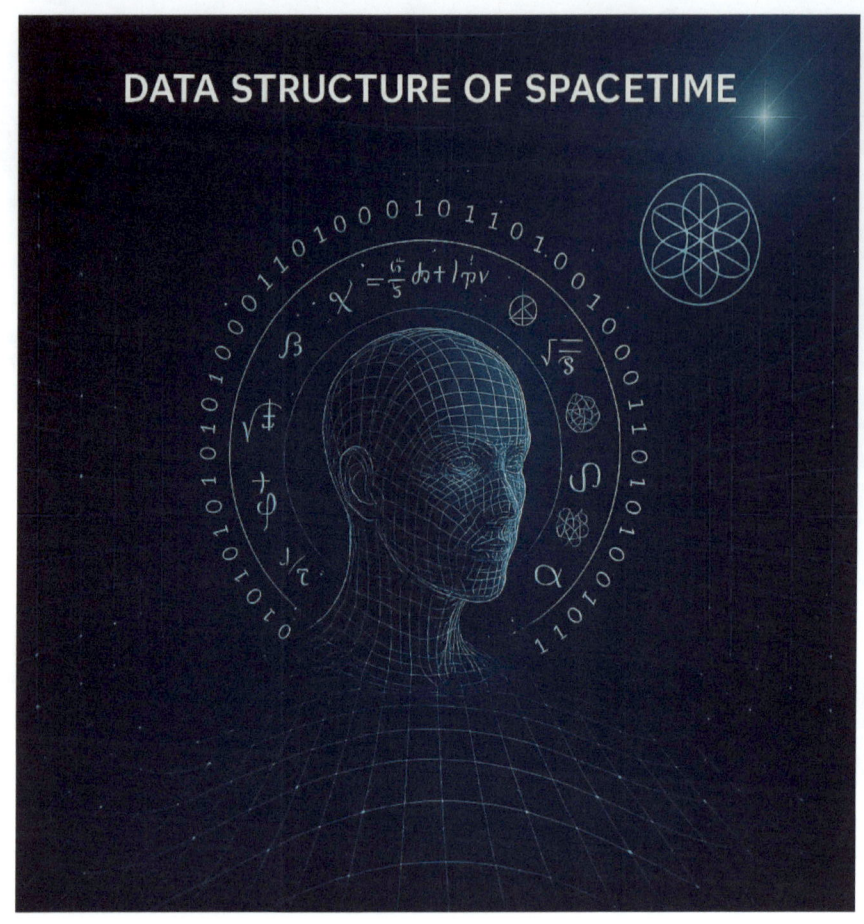

Diagram A2: *Data Structure of Spacetime*
Associated Chapter: Chapter Five – *The Hidden Dimensions of Reality*
This concept aligns with my core thesis that reality is not fixed but is constructed by consciousness

interacting with a deeper, data-like structure of spacetime.

Diagram Description:

1. A grid-like **fabric of spacetime** flows across the background—this resembles a mesh made of luminous points (like nodes) and thin connecting lines (data streams).

2. In the center, a **3D wireframe of a human head** is surrounded by rings of binary code, equations, and symbols.

3. From the head, **consciousness beams (light) ripple outward**, intersecting with the spacetime grid—each touchpoint collapses the grid slightly, **forming localized "realities."**

4. Floating above or to the side, small **symbols of sacred geometry**, Quranic calligraphy, or subtle DNA strands show that **spiritual and biological systems** are part of this matrix.

5. A soft glow of divine light at the top symbolizes **higher dimensional input** into the data structure (symbolizing Divine Will).

Narration:

This diagram illustrates supports my theory that spacetime operates like a multidimensional data structure—an informational matrix encoded with divine logic. It is not the fundamental substance of the universe, but an interface shaped by consciousness. As conscious agents (humans, angels, higher beings) interact with this structure, wave functions collapse, forming measurable reality.

Every point of awareness becomes a node, forming or interpreting data. These intersections determine not only physical phenomena but also metaphysical experiences, dreams, intuition, and prayer responses. In this view, **reality is rendered**, not merely observed. This diagram reflects the belief that divine order is woven into the unseen structure of the cosmos, and that consciousness is both a receiver and a transmitter of encoded divine equations.

It also reminds us that spiritual practices—such as dhikr, meditation, and prayer—are not merely rituals but acts of **programming alignment**, tuning our personal consciousness into harmony with the Most High's source code.

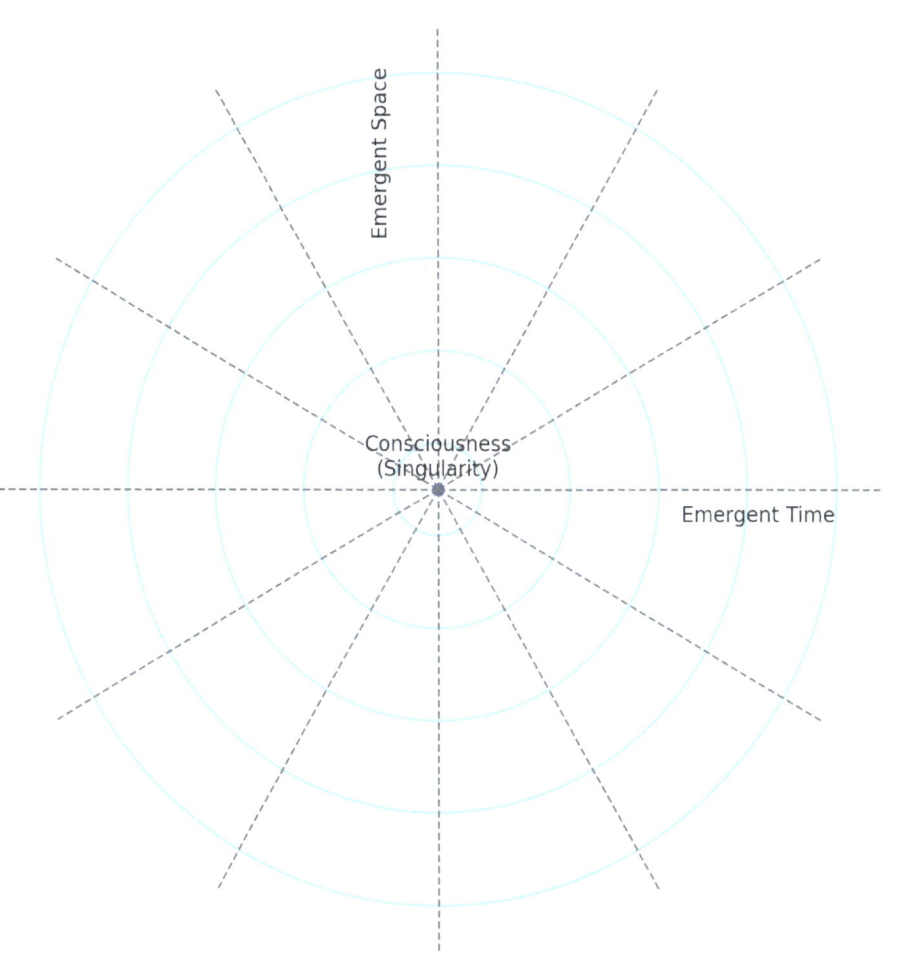

Diagram A3: The Amplituhedron and Consciousness

Associated Chapter: Chapter Eight – *The Equation of Time*

Diagram Description:

This intricate digital illustration explores the concept of the amplituhedron—a multidimensional geometric structure used in advanced physics to simplify particle interactions—and reimagines it through my lens as a metaphor for consciousness as the source of time and space.

1. **Central Structure: The Amplituhedron**

Dominating the center of the image is a radiant, multifaceted geometric shape—the amplituhedron itself. Its translucent surfaces gleam with interwoven gradients of gold, sapphire, and deep violet, evoking both mathematical elegance and spiritual resonance. It is depicted in semi-3D, hovering in space like a crystalline diamond, yet with dozens of subtle faces folding into one another, representing higher-dimensional logic collapsed into observable form.

2. **Emergent Waves of Time and Space**

Flowing outward from the amplituhedron's surface are ribbons of light and waves, moving in smooth spirals. These streams symbolize time and space emerging *from* consciousness rather than being containers of it. They are labeled subtly as "Temporal Flow" and "Spatial Web," demonstrating my belief that time is a created dimension, not an eternal constant.

3. **Overlay of Sacred Geometry and Grid Lines**

Superimposed on the amplituhedron are faint overlays of sacred geometric forms—Metatron's

Cube, the Flower of Life, and golden ratio spirals—linking this advanced mathematical object with timeless spiritual symbols. These designs are integrated with precision, illustrating the mathematical language of divine architecture.

4. Surrounding Field: Consciousness Grid

Around the main shape is a soft matrix or "grid" of glowing nodes connected by lines, extending in all directions. These represent conscious agents—individual points of awareness—interacting across dimensions. The grid pulses with vibrational energy, suggesting an intelligent fabric of consciousness that supports and interacts with the amplituhedron itself.

5. Color and Light

The palette uses a divine spectrum: deep violets and indigos to signify higher consciousness, soft golds to convey divine intelligence, and teal green for vibrational balance. Light emanates from the amplituhedron like a starburst, representing divine insight or revelation.

6. Symbolic Contrast: Internal vs. External Perception

One half of the amplituhedron glows from within, representing internal awareness and intuition. The other side is lit by a light source outside the frame, symbolizing external stimuli and experience. This dual illumination reflects the core idea that consciousness is both inward and expansive, birthing and interpreting reality simultaneously.

Narration: This diagram explores the idea that time and space are not fundamental but **emerge from deeper layers of consciousness and mathematical geometry.** The Amplituhedron, discovered in theoretical physics, suggests that certain geometric forms may underline all particle interactions. In this book's context, it becomes a metaphor for the divine mind shaping reality outside linear time—where consciousness becomes the stage upon which matter, and time arise.

Entropy (*Entropic time*)

$$E(t\gamma) = \lim_{\gamma \to 0} \left| \frac{1}{1 - ke^{-\gamma t}} \right|$$

Note at least two types of entropy: Entropic time, meaning that entropy is not a constant to what we experience in spacetime, where entropy can be considered the arrow of time that doesn't decrease but ever increases. Although it is hypothesized that as the dimensions progress higher for example: the 5th to the 11th dimension, relative to spacetime (4d Space), entropic time approaches to that of a constant which is zero change; meaning no decrease or increase of energy. Consequently, when there is no space and time; entropy is a constant (zero change).

Entropic time = E

γ = energy
t = time

Nonentropic time

Entropic time Plot

$$\lim_{d \to 30} \left(1 - \frac{1}{1} - 1 e^{-d}\right) = 1.$$

Entropic time / Nonentropic time axis

At D = 3 dimensions where entropy is not a constant.
Entropic time

$E(D)$

As the dimension progresses higher…

At D > 3 dimensions when entropy becomes a constant. Nonentropic time.

D = 3 to "n" dimensions

(d from 27 to 33)

Entropic time curve vs number of dimensions

27 28 29 30 31 32 33

Note: The science behind how it is possible for some people to detect ones that left this plane of activity (4D space) to a higher plane of activity (a fifth dimensional expanse). The bio magnetic fields that your heart produces I suspect disturbs the space time around the individual not unlike gravitational waves that opens portals to where conscious agents can enter and directly communicate with that individual if the individual's pineal gland is tuned to the gravitational waves that are converted to electrical impulses. These impulses converted from the vibrations of space time that the frequency is generated from, the individual can interpretate these electrical impulse as a language where only the individual who receives them will understand.

Diagram A4: Entropic Time Equation

Associated Chapter: Chapter Eight – *The Equation of Time and on Book Cover*

Diagram Description

This original diagram reflects my exploration of Entropic Time as a principle linking thermodynamics, dimensional theory, and divine cosmology. The design models how entropy—typically understood as disorder or the arrow of time—behaves differently depending on dimensional context.

Core Equation:

$$E(t\gamma) = \lim_{\gamma \to 0} \left| \frac{1}{1 - ke^{-\gamma t}} \right|$$

In this formula, entropy E is evaluated as a function of time t, a decay factor γ, and a dimensional constant k. As γ approaches 0, symbolizing higher dimensions where energy no longer decays, the equation resolves into a constant. This represents the idea that in elevated dimensions, entropy ceases to fluctuate—it becomes timeless.

Entropy Curve:

The diagram presents a curved entropy line (E) rising in 4D space and gradually leveling as dimensions increase.

- At D = 3 (our current dimension): Entropy increases—this is what we experience as the arrow of time.
- As D → 5 through 11: The curve flattens. Entropy

stabilizes, signaling a shift toward timeless equilibrium.

Dimensional Ascension Axis:
   The x-axis spans from 3D to 11D, suggesting a progression beyond conventional spacetime. Labeled transition points show where time begins to lose its arrow-like quality and approaches constancy.
- A subtle energy wave runs parallel to the curve, reinforcing vibrational resonance across dimensions.

Visual Style:
   The diagram is rendered with clean gradients—deep blue in the lower left fading into soft violet and radiant white in the upper right. These hues symbolize the shift from temporality to transcendence. A crystalline backdrop evokes divine stillness and higher-order symmetry.

   Narration: This diagram captures the essence of my entropic time equation. I believe entropy is not fixed—it changes with dimensional context. In our world, entropy always increases, guiding our sense of time from past to future. But as we ascend into higher dimensions—beyond the fourth and into the fifth and beyond—entropy stabilizes.

What we experience as time's arrow disappears. Instead of change and decay, we enter a state of dimensional equilibrium, where energy no longer increases or decreases. That is the state I call nonentropic time—a still, ever-present flow.

In such realms, time is not linear. It's not even sequential. It simply is—a perfect constant. I don't see this as emptiness. I see it as fullness—an ever-abundant field of potential, untouched by decay. This model is one way I bridge science and spirit, using the language of mathematics to describe divine realities.

Author's Note & Disclaimer: The Entropic Time Equation and its corresponding diagram represent my personal metaphysical hypothesis, emerging from a blend of scientific curiosity and spiritual insight. While it draws on elements of classical thermodynamics and higher-dimensional theory, this is not a mainstream physics model. It is offered as a philosophical tool to expand our understanding of time, energy, and the eternal nature of the Most High.

SUBCONSCIOUS MIND & INTERACTION WITH HIGHER DIMENSIONS

Diagram A5: The Subconscious Mind & Interaction with Higher Dimensions

Associated Chapter: Chapter Nine – *Conclusion & Reflection*

Diagram Description:

1. Central Figure

A human figure is depicted in seated meditation, centered in the image. The figure radiates a soft internal glow, symbolizing inner awareness and spiritual receptivity.

2. Concentric Spheres

Surrounding the figure are multiple translucent, concentric spheres, each representing a higher level of consciousness. These spheres grow more ethereal and abstract as they expand outward, illustrating the journey from the subconscious to the superconscious.

3. Toroidal Energy Fields

Flowing around the figure are dynamic toroidal shapes—doughnut-like energy fields—rendered in hues of indigo, emerald, and violet. These represent the vibrational resonance of the body and soul, as well as the continuous loop between inner awareness and universal consciousness.

4. Ascending Spiral Helix

Rising from the figure's crown is a luminous spiral, symbolizing the ascent of consciousness. This helix reaches toward a complex, multidimensional geometric structure above the figure.

5. Higher-Dimensional Geometry

At the apex of the spiral is a tesseract-like structure—an abstract four-dimensional cube—symbolizing the threshold into higher realms. It is inscribed with sacred symbols including:

    1. The number 7 (divine order)
    2. The number 19 (a mathematical key referenced in the Quran)
    3. Flowing script in Arabic and Hebrew, suggesting unity in divine revelation

6. Subconscious Realm

Below the figure, a dark, starlit basin represents the subconscious mind. Within this pool archetypal images appear:

    1. A feather (sensitivity and spiritual attune)
    2. A key (access to divine truth)
    3. An archway (portal to deeper dimensions)

7. Orbs of Light / Conscious Agents

Floating within and around the spheres are radiant orbs—representing conscious agents such as angels, ancestors, or higher-dimensional beings. These figures appear to move between the spheres, indicating that spiritual guidance flows through vibrational resonance.

8. Sacred Arc Frame

The entire scene is encircled by an arc inscribed with sacred numerals and divine script. This frame evokes the harmony of divine mathematics and the unity of knowledge across traditions and dimensions.

Narration: This visual portrays the human consciousness as a layered system, where the **subconscious acts as a bridge** between the physical self and higher-dimensional realms. Just as quantum fields operate unseen yet influence reality, the subconscious mind receives impressions from **spiritual agents**, ancestral memory, and divine intuition. This diagram affirms that by aligning our internal frequency—through dhikr, prayer, and awareness—we can access wisdom and guidance from realms beyond time and space.

ABOUT THE AUTHOR

I am Tyrone James, a truth-seeker, a questioner of convention, and a student of the unseen. My journey has been one of weaving together threads of mathematics, physics, and metaphysical exploration to better understand the magnificent harmony of creation. For me, equations are not merely symbols, they are the fingerprints of the Most High etched into the fabric of existence.

I wrote this book not as a theologian, nor a traditional scientist, but as one who has spent a lifetime asking deeper questions: What lies beyond the veil of our senses? How do numbers reveal divine patterns? Can science and spirituality, long seen as separate paths, actually lead us to the same Source?

In this work, I invite readers to journey with me across the realms of time, consciousness, and vibration—to explore how the divine reveals itself not only in sacred texts, but in the structure of the cosmos itself. It is my hope that this book serves as a bridge between faith and reason, heart and mind, and ultimately between humanity and the Divine.

NOTES

NOTES

NOTES

www.ingramcontent.com/pod-product-compliance
Lightning Source LLC
Chambersburg PA
CBHW041630220426
43665CB00001B/8